Material Flow Analysis

Edited by Sanjeev Kumar

Published in London, United Kingdom

IntechOpen

Supporting open minds since 2005

Material Flow Analysis
http://dx.doi.org/10.5772/intechopen.87885
Edited by Sanjeev Kumar

Contributors
Bir Bahadur Sherpa, Reetu Rani, Noor Irinah Irinah Omar, Sanjeev Kumar, Erwin Povoden-Karadeniz, Muñoz Ayala Israel, Vera Roberto Carlos, Qiang Li, Abdmouleh Makki, Ilyes Jedidi, Motohiro Yamada, Toshiaki Yasui, Masahiro Fukumoto

First published in London, United Kingdom, 2021 by IntechOpen
IntechOpen is the global imprint of INTECHOPEN LIMITED, registered in England and Wales, registration number: 11086078, 5 Princes Gate Court, London, SW7 2QJ, United Kingdom
Printed in Croatia

British Library Cataloguing-in-Publication Data
A catalogue record for this book is available from the British Library

Additional hard and PDF copies can be obtained from orders@intechopen.com

Material Flow Analysis
Edited by Sanjeev Kumar
p. cm.
Print ISBN 978-1-83962-956-3
Online ISBN 978-1-83962-957-0
eBook (PDF) ISBN 978-1-83962-964-8

For EU product safety concerns:
IN TECH d.o.o., Prolaz Marije Krucifikse Kozulić 3, 51000 Rijeka, Croatia, info@intechopen.com or visit our website at intechopen.com.

Meet the editor

Dr. Sanjeev Kumar has been working as an assistant professor in the Department of Mechanical Engineering at the National Institute of Technology, Raipur, India, since August 2021. Prior to this position, he served as Senior Researcher at CDL-IPE, Technical University of Vienna, Austria, from June 2018 to July 2021. Dr, Kumar also served as a Postdoc Researcher on the FWF project at IMAT, TU Graz, Austria, from May 2017–April 2018. He obtained his Ph.D. and MTech degrees in Welding Metallurgy from IIT Roorkee, India, while also obtaining a BTech in Mechanical Engineering from AKGEC-GZB UPTU, Lucknow, India. His primary research interest is welding, thermomechanical processing, and material design. He has published one book and more than thirty-eight research papers in recognized international journals, conferences, and books.

Contents

Preface

Material flow analysis has been a crucial topic in advanced metal and composites manufacturing processes, specifically for the processes of rolling, forging, extrusion, and welding that governs the quality of products through plastic flow behavior of different materials. These materials include new high/ultra-high-strength steels, and other lightweight alloyed metals and composites, such as those used in aerospace, automotive industries, biomedical engineering, and structural applications. The materials are used to reduce the cost and weight of materials; however, they increase quality and compatibility with human life. In the era of increasing importance to sustainable manufacturing, strengthening and joining by plastic deformation will find more application in various industries. These processes will help to develop lightweight products with compact design and desired directional properties, thereby helping in reducing the material and energy consumption. Usually, constitutive analysis of the flow behaviors yields assessable information on the relation between the flow stress behaviors and the thermomechanical state variables, that is plastic strain, strain rate, and temperature, and axial load in tension/compression desired for processing of advanced alloyed materials. The best consolidation of mechanical and physical properties is only possible with the right sequence of thermomechanical state variables. Furthermore, it is well known that complicated, large size, and compatible design of products is not possible to develop in a single-step manufacturing process. The joining process plays an important role to join similar/dissimilar metals during the welding process. In the last three to four decades, the solid-state welding process, e.g. friction stir welding and explosive welding, has been observed for high-ductility alloys and bimetallic plates and has been found to be the most important technology for lack of melting, less deformation, and fewer defects as compared to conventional fusion welding processes. In these processes, the high-speed stirring of the tool during friction stir welding and the collision of one plate with another plate during high velocity during explosive welding cause the joining area to experience severe plastic deformation. It may also be responsible for the degradation of mechanical properties and may also be beneficial. Thus, it becomes necessary to control their parameters, which can provide the appropriate quality.

Currently, the demand for composites is increasing due to the requirement of multiple environment (e.g., chemical, thermal, physical, and mechanical), and those composites can be the combination of metal/plastic, metal/polyethylene, metal/ceramic coating, etc. Apart from all these topics, liquid-liquid extraction ion-exchange or partitioning of organic compounds is exclusively used for extraction and purification of uranium, titanium, zirconium, etc.

Therefore, the right combination of thermomechanical, welding, composites, and the extraction process parameters may be favorable to reduce cost and longtime stability and may be compatible with human life. Thus, the aim of this book was a review of recent progress in the above-mentioned topics. This book includes six chapters. In the area of plastic deformation in materials, the topics covered are forming, joining, explosive welding, and directional mechanical properties for different grades of materials. While in the area of composites and extractions, the topics are covered on composites and the new extraction process.

Dr. Sanjeev Kumar
Assistant Professor,
Department of Mechanical Engineering,
National Institute of Technology Raipur,
Raipur, Chhattisgarh, India

Section 1

Plastic Deformation in Materials

Chapter 1

Plastic Deformation Behavior in Steels during Metal Forming Processes: A Review

Sanjeev Kumar and Erwin Povoden-Karadeniz

Abstract

The plastic deformation occurs in steels during metal forming processing such as rolling, forging, high-pressure torsion, etc. which modify mechanical properties of materials through the grain refinement, and the shape change of objects. Several phenomena in the scope of plastic deformation, such as hardening, recovery, and recrystallization are of great importance in designing thermomechanical processing. During the last decades, a focus of research groups has been devoted particularly to the field of metals processing of steel parts through plastic deformation combined with specific heat treatment conditions. In this review chapter, the current status of research work on the role of plastic deformation during manufacturing is illuminated.

Keywords: Plasticity, Ferrous metal, steel, SPD, Deformation, Strengthening, Flow Curves

1. Introduction

In the modern era, the demand for lightweight material products is being increased in industries e.g. aerospace [1, 2], automobiles [3, 4], buildings [5, 6], trains [7, 8], forged connecting rods and pistons [9], bridges [10], naval [11–13], etc. for a high living standard (see **Figure 1**). The researchers are dedicating high effort to increase the strength to weight ratio by grain refinement through applying heat treatments [14–20], mechanical processing [21, 22], and a combination of both i.e. thermomechanical processing (TMP) [23–27]. TMP methods are being used in the manufacturing unit to fulfill requirements of grain refinement of materials and create optimum semi-finished and finished products for the applications. The grain size of steels is an important factor that affects all aspects of the mechanical, chemical, and physical behavior of metals to the surrounding media. It is well known that the smaller grains support an increase in grain boundaries in the matrix. In particular, according to the Hall-patch law, the reduction in grain size improves material properties like strength Eq. (1), hardness, and impact toughness except for the ductility of steels [28].

$$\sigma_Y = \sigma_i + K_Y / \sqrt{D} \tag{1}$$

Where: σ_i = friction stress, D = grain diameter, K_Y = yield coefficient or "locking parameter" that shows the relative hardening contribution of grain boundaries.

Figure 1.
Some important examples of industrial applications which developed using metal forming processes.

Some of the major metal processing steps are often involved such as rolling, forging methods with wide temperature ranges (cold, warm and hot deformation temperature ranges) for the grain refinements [26, 27, 29–32]. The high-pressure torsion, equal channel angular pressing (ECAP), direct/indirect extrusion methods etc. are being used for ultrafine grains in which plastic transformation reaches over strain 1 through severe plastic deformation (SPD) [33–35]. In this SPD processes, the large shear stress involved usually results in a complex stress state resulting in a high defect density and homogeneous ultrafine grains.

During metal forming processing, the steel experiences different metallurgical phenomena like work hardening, dynamic recovery, dynamic recrystallization, flow instabilities, etc. [32, 36–38]. The effect of these metallurgical phenomena can be understood through the interpretation of flow curves [26, 31, 39, 40]. Where, the flow stress dependent on various processing parameters such as temperature, strain rate, and strain, etc. that can typically been described via constitutive equation.

This chapter focuses on plastic deformation behavior which can be controlled through processing parameters that affect microstructure refinement and associated mechanical properties of metals and steels during forming.

2. Some common metal processing and joining setups

A schematic diagram of rolling, forging, and high-pressure torsion (HPT) is shown in **Figure 2**. These represent basic processes within metals engineering, which change the shape and microstructure through plastic deformation for different products and applications. **Figure 2a** illustrates the rolling process setup, where the billet is pulled out between pairs of rollers which reduce the thickness of plates and grain size as well as defects like porosity and inclusions of the billet. The resulting refined grains are found to be elongated along the longitudinal direction. In the forging process, the force is imposed on objects either by hammer and anvil or in a large forging tool (called drop hammer) which results in desired and controlled shape changes (**Figure 2b**). Eventually, the HPT setup is one kind of torsion process in

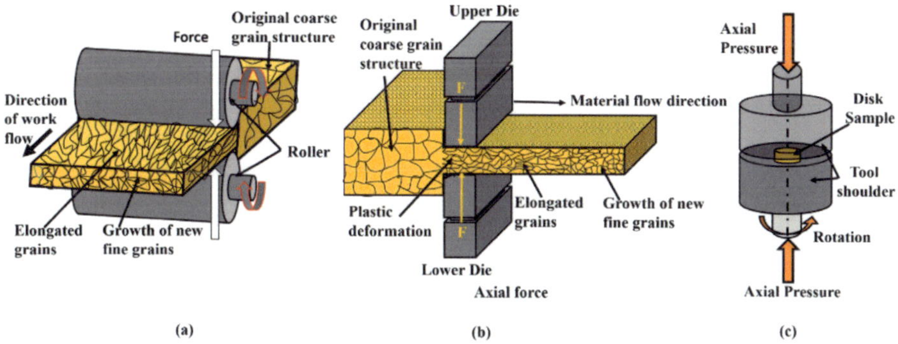

Figure 2.
Schematic diagram of metal forming processes (a) forging, (b) rolling, (c) high-pressure torsion [41].

Hot compression testing	
Strain rate (1/s)	**Machine details**
10^{-3} to <5	Basic unit, Gleeble
10^{-2} to 100 <3x10^{2}	Hydrawedge unit standard specimen ϕ10 x L15 mm modified specimen ϕ10 x L0.5 mm
0.1 to 5x10^{2}	Cam Plastometer and Drop Test
2x10^{2} to 10^{4}	Split Hopkinson Pressure Bar
10^{3} to 10^{5}	Taylor impact machine
More than 105	Gas gun (single & two-stage)
Torsion/Multiaxial/Shear testing	
<10^{-1}	Conventional shear machine
10^{-1} to 10^{2}	Hydrawedge unit, Gleeble
10^{-1} to 10^{3}	Torsion unit, Gleeble
10^{2} to 10^{4}	Split Hopkinson Pressure Bar
10^{3} to 10^{4}	Double-notch Shear and Punch
10^{4} to 10^{7}	Pressure-shear plate impact machine

Table 1.
Wide range of hot compression test setups [42].

which material undergoes severe plastic deformation via applying both compressive force and twisting action concurrently under high pressure (**Figure 2c**). The sample for SPD is located between two anvils, where the top anvil provides a compressive force on the sample while the bottom side anvil rotates along on axis. This setup generates shear strains in the object which are responsible for the development of ultrafine grains. Therefore these setups can be supportive of favorable mechanical properties and good product performance.

Important experimental machines, being used for a wide range of strain rates are listed in **Table 1**. In this list, the Gleeble machine can be used for axial compression testing with the strain rate between 0.001 to 100 for standard samples with a diameter of 10 mm and a length of 15 mm. It should be noted that much higher strain rates up to 3000 are feasible when a shorter sample, typically less than 1 mm, is chosen. A wide range of strain rates can be achieved using other compression testing machines, Cam plastometer, Slip Hopkin, Taylor, and gas gun machine.

Some important torsion test setups are listed for shear testing with a wide range of strain rates within the framework of SPD. All of the listed setups are supportive for controlled and taylored TMP in order to achieve an optimized balance of processing costs, time, and materials properties for various industrial applications.

3. A basic understanding of microstructure

The morphology of materials can be defined through shape, size, and structure that plays an important role in both mechanical and corrosion resistance properties. It is well known that all materials are composed of atoms that are arranged in short/long-range order with regular/irregular patterns, those solids are familiar as crystalline and non-crystalline, respectively. The crystalline metals with different crystal structures, such as body-centered cubic, face-centered cubic, or hexagonally closed packed, are prorated into the single crystal and polycrystalline categories. Conversely, most polycrystalline metals are composed of a collection of many small single crystals named grains and are similar to pomegranate fruit, which is made up of many small seeds (see **Figure 3a**). The grains are separated from each other by grain boundaries while preserving the integrity of the metal. Similarly **Figure 3b** shows one grain (shown by yellow dotted line) that has a subgroup of several laths, and every lath having several crystal atoms.

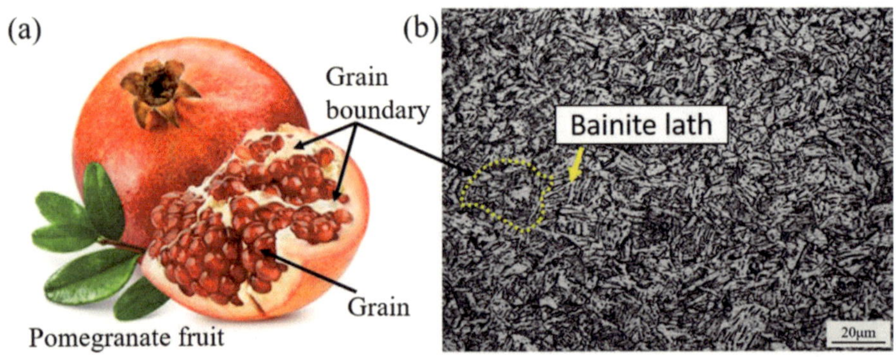

Figure 3.
The photographs of (a) pomegranate fruit, which compound of grains and separated by grain boundary like metal structure, (b) high strength steel structure consists of bainitic ferrite and martensite [27].

4. Some basics of plastic deformation mechanism

We know that the plastic deformation permanently changes the dimension and shape of metal, whereas in terms of microstructural changes only the number density of dislocations increases, whereas crystal structures including lattice parameters of metals typically remain unchanged. Slip and twinning processes, which are shown in a simple model presentation in **Figure 4**, are responsible for this macroscopic change of shape and dimensions. Slip implicates sliding of abutting blocks of a crystal along definite crystallographic planes, called slip planes. A slip occurs when shear stress applied to the material exceeds a critical value. During slip, each atom usually moves the same integral number of atomic distances along the slip plane producing a step, without change of the crystal orientation (**Figure 4b**). Grain boundaries represent obstacles for the slip movement as the slip direction, according to **Figure 4a**, will be usually changed across the boundary. This implies

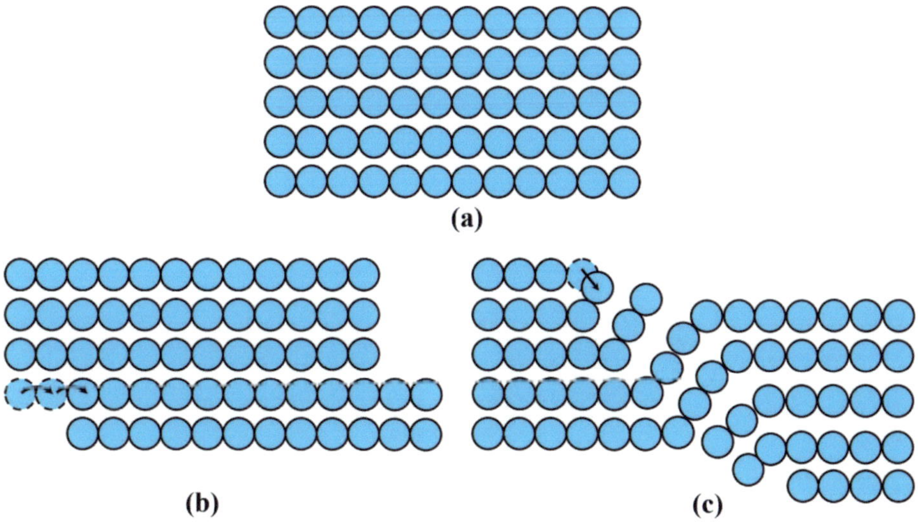

Figure 4.
Schematic representation of slip and twinning mechanisms in metals during plastic deformation (a) original position of atoms within a crystal lattice, (b) atoms movement by slip, (c) atoms movement by twinning [43].

that the strength of polycrystalline materials will be higher than that of a single crystal of the same material.

In twinning, each atom moves by only a fraction of an interatomic distance relative to its neighboring atoms (see **Figure 4c**). The twinned portion of the crystal is a mirror image of the parent crystal. The orientation of the twinned region is however different from the untwinned region (**Figure 4a**).

5. Terminology and summary of TMP related mechanisms

Metallurgical incidents during the TMP may act statically or dynamically on the material. This depends upon the rate of load and temperature conditions and strongly affects grain refinement. Microstructural evolution during TMP largely depends on the ability of dislocation movement during plastic deformation, which has consequently also a considerable impact on the mechanical properties of materials. The terminology of several mechanisms related to TMP is introduced in the following. These can be understood with the help of the flow stress–strain diagram (**Figure 5**) interpretation [44]. The flow stresses σ_c, σ_p, and σ_s mean the critical, peak, and steady state conditions, respectively. The combined effect of work hardening (WH) and softening mechanisms on flow curves are categorized into distinct regions: I) hardening, II) critical, III) softening and IV) steady-state. WH and dynamic recovery (DRV) occur in the first region where WH dominates and flow stress rises steeply. The second region is the critical zone where DRV and WH both are decreased and new dynamic recrystallization (DRX) initiates. Subsequently, DRX is clearly observed in the third region associated with softening. The fourth region is a steady-state where only DRX occurs. Key mechanisms and their terminology are treated in detail in the following.

5.1 Work hardening

Work hardening (WH) is also called strain hardening or cold hardening. It is the process of making a metal stronger and harder below its recrystallization

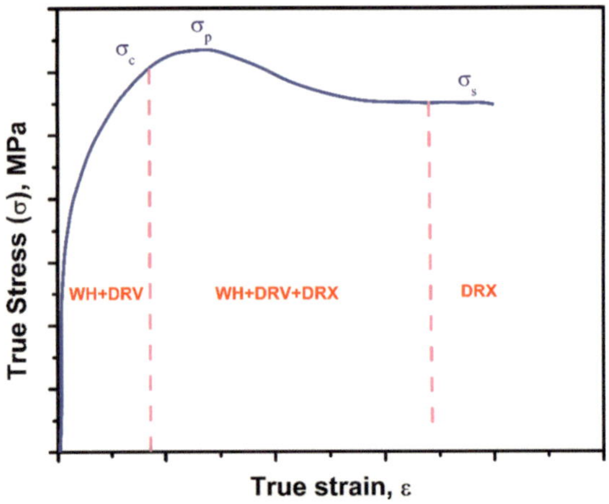

Figure 5.
Schematic flow stress–strain diagram [44].

temperature by increasing dislocation density via plastic deformation. Dislocations will be pinned by each other. Also, as a consequence this highly "faulted" microstructure will prevent the propagation of cracks. With increasing the temperature, the chance of rearrangement of matter and also dislocations is higher which contributes to lower strength at increased ductility.

5.2 Recovery

Recovery is a softening process that refers to the relieve of part of the internal energy stored within the microstructure, taking place before recrystallization in a deformed material. It normally occurs above the recrystallization temperature where the movement of atoms, i.e. the atomic mobilities and derived diffusion is considerably facilitated. Diffusion increases rapidly with rising temperatures and tends to recover strained regions to the "original" unstrained structure (**Figure 6a**). The extent of recovery depends, among other parameters, on the stacking fault energy (SFE), the type and amount of solute atoms of the material, particularly in the context of dislocation dissociations, which determine the rate of dislocation climb and cross slip. In low SFE metals, recovery as well as cross slip and climb of dislocation is difficult, while the climb is rapid and significant recovery may occur in metals and alloys with a high SFE [46].

Two types of recovery are known, static and dynamic recovery. Static recovery (SRV) occurs at high strain rates where jerky microstructural response of dislocation dynamics prevails. Technologically, this is the case for instance during friction stir welding (FSW) and other torsion processing. Dynamic recovery (DRV) occurs at slower strain rates where thermal activation of the metastable positions within the dislocation structure leads to steady-state during metal processing e.g. hot-rolling, extrusion, and forging processes, It is commonly accepted that both DRV and SRV reduce the stresses through changes in dislocation structure due to subgrain growth, dislocation annihilation, and dislocation rearrangement into lower-energy configurations (such as planar dislocation boundaries). Overall, ductility is improved by recovery, while the strength of materials is reduced [47].

5.3 Recrystallization

The recrystallization associates with the nucleation of new strain-free grains and their subsequent growth in deformed microstructure when internal energy reaches a critical value (**Figure 6b** and **c**). When the recrystallization process arises during

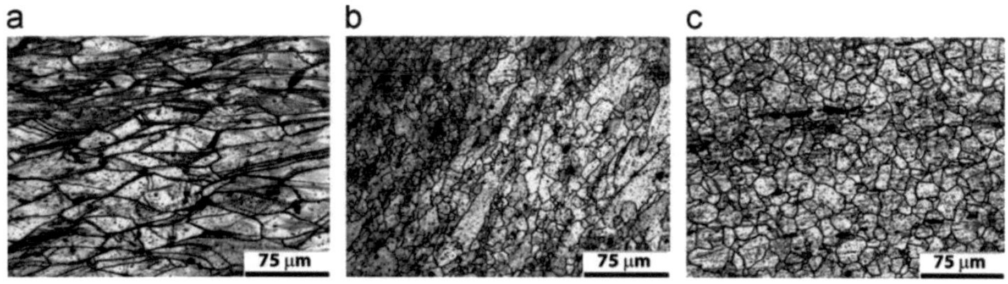

Figure 6.
The optical microstructure of deformed stainless steel samples: (a) recovered grains (b) partial recrystallization (c) area of full recrystallization [45].

Type of process	Mechanism	Materials type
Hot deformation ($T > 0.5T_m$)	dDRX	Category L & M
	cDRX and DRV	Category H
Cold/warm deformation ($T < 0.5T_m$)	cDRX	All Categories
Hot torsion ($T > 0.5T_m$), other SPD processes	DRV + dDRX+cDRX	All Categories

Where,
Low and medium-range SFE materials (*Category L & M*): Copper, Gold, Lead, γ-iron, Ni and their alloys.
High SFE materials (*Category H*): Aluminum, Magnesium, α-iron, and their alloys.

Table 2.
The details of materials and acting mechanisms during the hot deformation process [34, 49, 50].

deformation processes, it is called dynamic recrystallization (DRX). In contrast, when it takes place after deformation or during post-processing like the annealing process, it is known as static recrystallization (SRX) [48]. When DRX is not completed within deformation, this is termed meta dynamic or post dynamic DRX (mDRX) [48]. Moreover, two types of DRX can be distinguished. In discontinuous DRX (dDRX) strain-free grains nucleate and grow rapidly, thus consuming the surrounding strain hardened matrix, while continuous DRX (cDRX) involves the generation of new grain boundaries by the continuous misorientation of nearby subgrains. The combined effect of cDRX and dDRX phenomena takes place during higher strain conditions which are possible during torsion, other severe plastic deformation processes [29]. Since the rate of annihilation due to dynamic recovery is not sufficient to complete with the strain hardening rate in low SFE materials, the dislocation density increases continuously in this case. Contrary, high SFE materials act in favor for higher mobility of dislocation, and consequently dynamic recovery becomes involved as an operating mechanism [49]. The details of materials and the type of possible phenomena are illustrated in **Table 2**.

The range of dDRX and cDRX can be understood through the schematic diagram between processing temperature and strain rate (see **Figure 7a**). dDRX phe nomena increase above the melting temperature (T_m) when the strain rate decreases while the cDRX phenomena decrease with decreases in processing temperature and increases in strain rate. cDRX occurs in all SFE materials [53] when the temperature falls below 0.5 T_m, however, the dDRX takes place only in low and medium-range SFE materials above 0.5 T_m wherein dynamic recovery is slow after accessing a critical strain value, as can be seen in **Figure 7a** [34, 54, 55]. The grain nucleation and growth during dDRX is the same as for primary DRX which occurs during heating in cold-worked materials. Localized nucleation and growth at local grain boundary

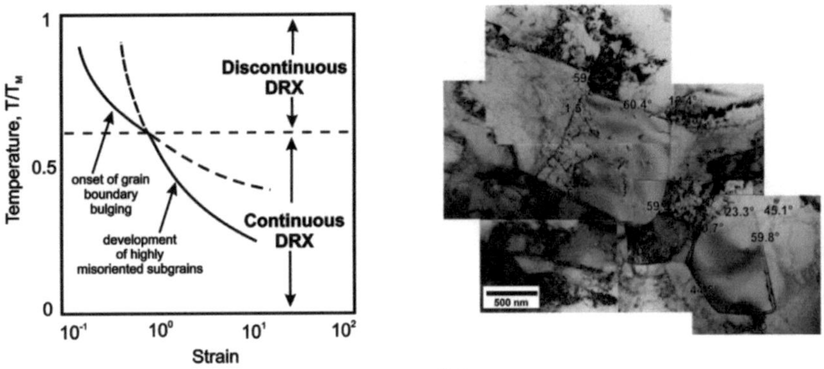

(a) (b)

Figure 7.
(a) Schematic correlation between cDRX and dDRX [51], (b) dDRX nuclei in austenitic stainless steel at 800°C with strain rate 0.001 s^{-1} [52].

bulging can be seen in **Figure 7b**. It is obvious that the dDRX nuclei contain a much lower dislocation density than the deformed region and these nuclei are distinct from highly disturbed substructures with twin boundaries and low angle dislocation sub boundaries.

One additional terminology has recently been denoted as post-DRX which occurs during the annealing process in deformed materials [55].

6. Impact of plastic deformation parameters on microstructure and properties evolution

Some major metal processing parameters such as temperature, strain rate, and strains that impact steel microstructures and their flow curves are illuminated in detail in the following.

6.1 The role of work-hardening rate

The WH rate enables strengthening and hardening to the materials below the recrystallization temperature. Rapid WH rates are realized in low strain regions due to increases in dislocation density while at later strain increase, the effect starts to decrease due to recrystallization of new strain-free grains [31, 36, 40, 50, 56]. In WH, dislocations are preferably pinned, which will impede crack propagation on the microscale. With increasing temperature, the probability of rearrangement of atoms is higher which assists lower strength but increases the ductility of materials. Samantaray et al. [36] have reported for 316 L stainless steel that the WH rate starts rapidly with increasing temperature and strain rate at a specific value of strain (see **Figure 8**). The WH rate gradually decreased at higher temperature with increasing strain while it falls more rapidly under lower temperature conditions.

Lin et al. [44] have derived the following model for the influence of dynamic recovery during WH (see Eq. (2)) and dynamic recrystallization Eq. (3) under different deformation conditions within TMP.

$$\sigma = [\sigma_{DRV}^2 + (\sigma_0^2 - \sigma_{DRV}^2)\exp(-\Omega\,\varepsilon)]^{0.5} \tag{2}$$

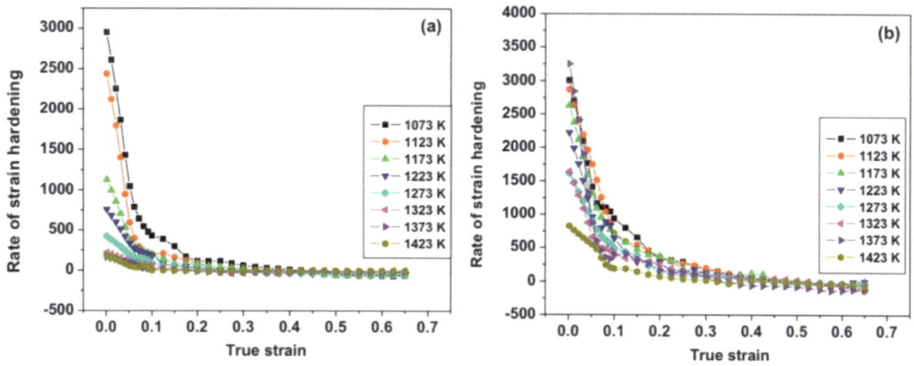

Figure 8.
The impact of temperature and strain rate on work hardening rate of stainless steel [57].

$$\sigma = \sigma_{DRV} - (\sigma_P - \sigma_{DRX})\left\{1 - exp\left[-K_d\left(\frac{\varepsilon - \varepsilon_c}{\varepsilon_P}\right)^{n_d}\right]\right\}\varepsilon \geq \varepsilon_c \tag{3}$$

Where σ is flow stress; σ_{DRV} is steady-state stress due to dynamic recovery; σ_{DRV} is steady-state stress due to dynamic recrystallization; σ_0 is yield stress; ε is strain; ε_P is peak strain; ε_c is critical strain; *Ω* is coefficient of dynamic recovery.

Generally, the critical strain acknowledgeable for the start of DRX can be calculated either by deformed microstructure or flow stress curves [58], In which flow stress curve analysis are simple and easier while microstructural are complicated. This flow curve analysis method was proposed in 1981 by Mecking et al. [59] and later it developed by Ryan et al. [60] and McQueen et al. [61] emphasize the point where DRX occurs on the flow curves. This method allows to find out the critical strain point where the flow curve changes due to the formation of new strain-free grains via DRX.

6.2 On properties derived from flow curves and relation to microstructures

The flow stress–strain curve reflects the changes in the material through plastic deformation during dynamic loading [25, 32, 62–66]. The flow stress can be influenced by several factors like chemical composition, crystal structure (e.g., steel matrix - bcc, fcc, Mg-base - hcp, and others) [50, 67], different phases and compounds [17, 30, 50, 68–70], grain boundaries [25, 50, 71, 72] as well as imperfections [34, 50, 55, 73, 74]. Other factors such as friction (σf), thermal (σt) and athermal (σa) terms also affect flow stresses, as indicative by relations in Eq. (4) [75].

$$\sigma = \sigma_f(\dot{\varepsilon},T) + \sigma_t(\dot{\varepsilon}, \varepsilon, T) + \sigma_a \tag{4}$$

Where T is temperature, $\dot{\varepsilon}$ is strain rate and ε is strain. σ_a represents the internal stress which occurs due to long range barriers to dislocation motion in the materials, while σ_f term reflects the stress needed to overcome the lattice friction depending on strain rate and temperature.

In addition, processing temperatures and strain rates are equally important for the plastic deformation behavior. Therefore, the dynamics of TMP can be understood through the investigation of microstructural changes combined with interpretations of trends of flow stress–strain curves which depend on DRV and DRX, and SRX [76, 77]. I.

It is noticed in most cases that flow stress decreases with increase in temperature and depends upon the applied strain rate [27, 32, 36, 78]. In terms of temperature, strain, and strain rates, the flow curves can be expressed by Eq. (5) [75].

$$\sigma = \frac{2}{\sqrt{3(1-m)}} K\varepsilon^{n}\, \dot{\varepsilon},^{m} \exp(-\beta T) \tag{5}$$

Where m stands for strain rate sensitivity, n represents the strain hardening exponent, and K, β represents material constant.

In the following, some flow curve trends of different steels and underlying phenomena are discussed.

Researchers reported that the series of flow curves are subjected to different temperatures and strain rates for different grades of steels [26, 27, 44, 45, 64]. Lin et al. [44] have reported interesting results for hot deformation of 42CrMo grade high strength steel in which they found that flow stress increased with decreasing temperature (**Figure 9a**) while it increased with strain rates (**Figure 9b**). At the slower strain rate in different ranges of temperatures, the flow stress will decrease with increase in temperature due to increase in the amount of cross slip screw dislocations and climb of edge dislocations, as well as vacancy diffusion. This results in an increase of grain boundary mobility and energy accumulation at boundaries for the nucleation and growth in DRX grains and dislocation annihilation which is responsible for the decrease in flow stress [27, 64].

Kumar et al. [27] have found for hot deformed condition in high strength steel that flow stress increases continuously at lower deformation temperature (750–850°C) due to continuation of work hardening phenomena High temperature showed the higher steady-state condition where DRX was dominant. While both DRV and DRX were dominant at all strain rates with decreasing temperature, the dDRX phenomena was more prominent at slow strain rate (0.001 s-1) at 900°C due to nucleation of unstrained grains that occurs normally in low SFE high strength steels. A flow curve without pronounced peak stress, but which exhibits

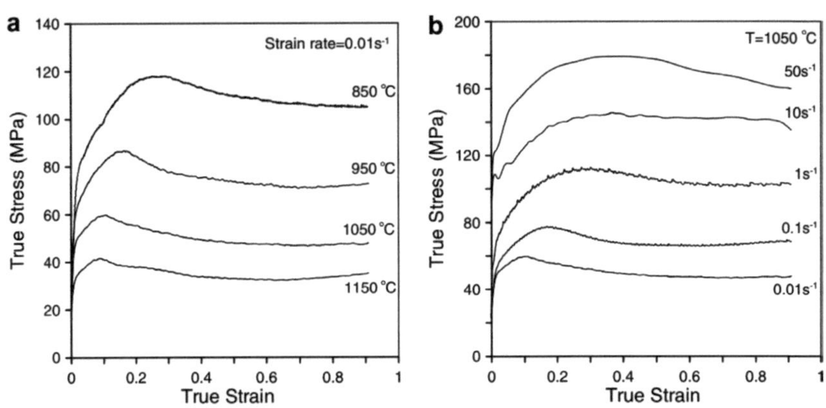

Figure 9.
The true stress–strain curves at different temperatures and strain rates for 42CrMo steel [44].

a steady-state, is generally associated with dynamic recovery being the dominant restoration mechanism [79].

Zhang et al. [31] have reported that several original grains were broken and recrystallized new grains showed up along the grain boundaries at the deformed condition at 900°C with a high strain rate 10 s^{-1}, which indicates that the deformed morphology is inhomogeneous. In contrast, at the same temperature but with a lower strain rate (1 s^{-1}), DRX was observed along the grain boundaries. This is due to local temperature rise within the samples during deformation. It is interesting to note that at the initial stage of strain, the flow stress increases steeply due to work hardening phenomena in materials having higher carbon content and less austenite stabilizing alloying elements. It reaches a peak value before going into the softening stage.

Souza et al. [45] have documented results of hot deformation testing at elevated temperatures with different strain rates in austenitic stainless steel. Also, some differences could be seen in the work hardening phenomena; the slopes of the flow stress curves changed. In the initial work hardening region, the increase of dislocation density during deformation is controlled by the competition between storage and annihilation of dislocations, i.e. opposing contributions of work hardening and the dynamic recovery due to the change of dislocation density with deformation.

7. SPD impacts on the structure and mechanical properties of steels

Severe plastic deformation where metal grains are heavily deformed is realized by using several setups of plastic deformations like high-pressure torsion, equal channel angle pressing, multi-axial forging, twist extrusion, accumulated roll bending, and constrained groove pressing [22, 34, 80]. Severe deformation produces not only a strong direct impact on the mechanical properties i.e. high strength, low-temperature toughness, superior plasticity, good ductility, and good wear resistance of high manganese grades steel but also on other important properties such as thermal stability, diffusion, radiation tolerance, and corrosion properties, which are indirectly associated with material stability and durability.

The high manganese steels (Mn) are advanced high strength austenitic steels that contain Mn between 3 to 31% wt. These steels are known as Hadfield steel, damping steel, complex steel, transformation induced plasticity steel (TRIP), and twinning induced plasticity steel (TWIP) [81, 82]. In all of these, Hadfield steel was firstly discovered in 1882 by Sir Robert Hadfield [83] while TWIP steel is one of the latest fully austenitic steel which is developed in the early 1990s by Japanese steelmakers Kobo steel, Nippon, and Sumitomo steel organizations.

It is well known that ultrafine and nanocrystalline structure depends on three mechanisms; martensitic transformation, dislocation motion, and twinning and twin evolution where stacking fault energy (SFE) of material plays an important role. **Figure 10** reflects the relation between strain-induced mechanism vs. temperature and SFE for Fe-20Mn-4Cr-0.5C steel. It shows that retained austenite can be converted into ε-martensite and strain-induced by a twinning mechanism at a lower temperature. Thus, the calculation of martensitic start temperature *and* SFE value is necessary to achieve the right combination of mechanical and other properties in low SFE high Mn steels. It is well known that the SFE of materials depends on chemical composition and on temperature [84–86]. The high Mn steels have a low SFE between 15 to 50 mJ/m2 [31, 85, 87].

Allain et al. [86] reported results for Fe-22Mn-0.06C steel where the temperature influences the SFE values and strain-induced mechanism, which can be seen in **Table 3**. The strain hardening and mechanical behavior of steels strongly depend

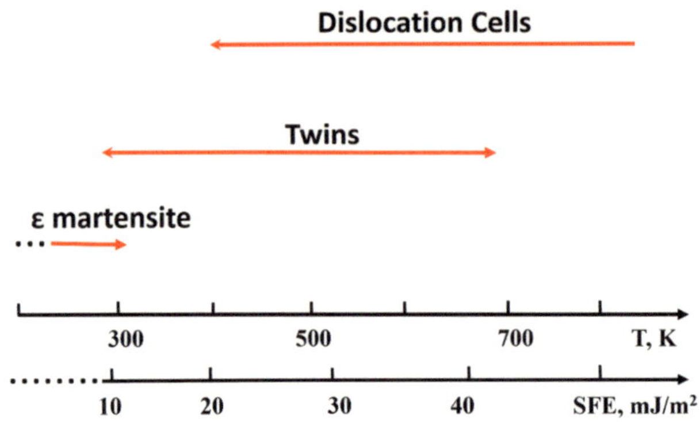

Figure 10.
Effect of SFE and temperature on deformation mechanism in Fe-20Mn-4Cr-0.5C steel [84].

Temperature, K	SFE value, mJ/m^2	Plasticity mechanism
77	10	Dislocation gliding
293	19	Dislocation gliding and twinning
673	80	Dislocation gliding and ε-martensitic transformation.

Table 3.
The deformation mechanism at different temperatures for Fe-22Mn-0.6C steel [86].

on the SFE, which is responsible for the activation energy of a deformation mechanism [87].

The mechanical properties of Hadfield Mn austenitic steels can be improved through high rate strain hardening where two phenomena (i.e. dislocation accumulation and twinning) act preferentially during plastic deformation [82]. This is attributed to the strain hardening transformation where austenite phase transforms into ε or α-martersite, and twinning, dynamic strain aging, dispute between dislocations with stacking faults occurs. In this connection, Yan et al. [88] have tried to improve the hardness values by shot pinning method, whereby hardness values could be increased with increasing in operation time. This was attributed to the increment in density of dislocations, dislocation accumulation, and formation of twinning. The influence of higher strain rate (between 10^3 to 10^5/s) attains great impact on mechanical behavior and wear resistance properties of high austenitic Mn steel which may be linked to dynamic strain aging and may delay fracture [81, 89–93].

Over the past few years, many researchers have reported work on TRIP and TWIP steels and achieved better mechanical properties by plastic deformation at high strains (more than 1) [22, 73, 74, 81, 89, 91, 94–96]. Both TRIP and TWIP steels are fully austenitic steels with less carbon content than hadfield steel. The initial microstructure of TRIP steel is consisted of martensite, bainite and ferrite with retained austenite. The fraction of carbon enriched retained austenite in TRIP steels is between 5 to 30% which transforms into martensite by displacive mechanism during SPD process. This behavior has attained great improvement in strength and toughness properties [96, 97].

A critical issue remains hydrogen embrittlement in TRIP steels, promoted by a displacive mechanism where the relevance of different solubility and diffusivity in the parent austenite has been discussed [96].

Sevsek et al. [90] reported the effect of strain rate on medium Mn X6MnAl12–3 steel. The softer austenite region was strained locally and transformed into

Figure 11.
Optical morphologies of Fe-22Mn-0.6C steel subjected to high strain deformation: (a) unstained; strained with (b) 18%, (c) 26%, and (d) 34% [81].

martensite which depended on the strain rate sensibility. The deformation induced phase transformation of austenite to martensite is partially suppressed at lower and higher strain rates. The impact of high strains in fully austenitic Fe-22Mn-0.6C steel is predominantly controlled by twinning plasticity mechanism (see **Figure 11**) as suggested by Jacob et al. [81]. The initial microstructure of Fe-22Mn-0.6C steel is a single-phase austenitic steel with few twinned grains (**Figure 11a**). They found that the fraction of twinning is increased with increasing strains (**Figure 11b–d**), where most of the internal energy was used for recrystallization and rest for grain growth [87]. They concluded that twin boundaries act as a hindrance to the dislocation glide providing work hardening effect.

In the same way, Kang et al. [98] have reported HPT tests for TWIP steel, in which they found that both stress and hardness values increased with an increase in the number of turns. This was related to grain refinement. It was also noticed that the inhomogeneity in morphology and volume of low and high angle grain boundaries increased with the increase in the number of turns, associated with higher stress and lower elongation. The hardness at tip location was found to be lower in all strain ranges due to the lower extent of plastic deformation while at the edge it was higher.

8. Concluding remarks

This review chapter focuses on plastic deformation behavior which can be controlled via processing parameters. Their optimisation is responsible for a refined microstructure, typically associated with beneficial mechanical properties of metals and steel due to metal forming. In other words, an appropriate combination of processing parameters enables one to fabricate products that will be defect-free on the microscale, which represents an important demand of customers. It is noticed that the flow stress increases with an increase in strain rate when the temperature is constant while it decreases with an increase in temperature when the strain rate is constant. The dDRX phenomena occur under axial stress hot deformation conditions while cDRX phenomena are linked to torsion deformation conditions during severe plastic deformation at relatively low temperatures. Plastic deformation acts differently in the case of high Mn austenitic TRIP steels where retained austenite is transformed into martensite by displacive mechanism and induced strain forms twinning which improves strength and toughness of steels. In contrast, the high-Mn fully austenitic steel such as TWIP steels generate huge amount of twinning structure by induced high strains and do not show phase transformation like TRIP steels.

Author details

Sanjeev Kumar[1*] and Erwin Povoden-Karadeniz[1,2]

1 Christian Doppler Laboratory for Interfaces and Precipitation Engineering CDL-IPE, Technische Universität Wien, Vienna, Austria

2 Institute of Materials Science and Technology, Technische Universität Wien, Vienna, Austria

*Address all correspondence to: sanjeev.kumar@tuwien.ac.at

References

[1] R. A. Anderson, M. W. Touma, G. Ashe, and J. F. Conlon, "Optimized Design Parameters for Welded Tmcp Steels," 1997.

[2] A. Forgas Júnior, J. Otubo, and R. Magnabosco, "Ferrite quantification methodologies for duplex stainless steel," J. Aerosp. Technol. Manag., vol. 8, no. 3, pp. 357-362, 2016, doi:10.5028/jatm.v8i3.653.

[3] A. Chatterjee, A. Moitra, A. K. Bhaduri, D. Chakrabarti, and R. Mitra, "Effect of heat treatment on ductile-brittle transition behaviour of 9Cr-1Mo steel," *Procedia Eng.*, vol. 86, no. July 2016, pp. 287-294, 2014, doi: 10.1016/j.proeng.2014.11.040.

[4] O. G. Comineli, "Investigations on the Influence of Oxidation in Causing Hot Cracking in Steels," in *Thermo-mechanical Simulation and Processing of Steels (SimPro '16), Ranchi, India*, 2016, no. Feb, pp. 579-588.

[5] Q. Xue, D. Benson, M. a. Meyers, V. F. Nesterenko, and E. a. Olevsky, "Constitutive response of welded HSLA 100 steel," Mater. Sci. Eng. A, vol. 354, no. 1-2, pp. 166-179, 2003, doi: 10.1016/S0921-5093(03)00007-8.

[6] S. Kumar, S. K. Nath, and V. Kumar, "Continuous cooling transformation behavior in the weld coarse grained heat affected zone and mechanical properties of Nb-microalloyed and HY85 steels," Mater. Des., vol. 90, pp. 177-184, 2016, doi: 10.1016/j.matdes.2015.10.071.

[7] T. Y. Hsu and X. Jin, *Advanced Steels*. 2011.

[8] K. Jármai and B. Bolló, *Vehicle and Automotive Engineering*. 2017.

[9] D. Niu, J. Zhang, P. Xiong, G. Hao, S. Liu, and W. Guo, "High temperature fatigue and oxidation characteristics of forged steel piston materials," *Eng. Fail. Anal.*, vol. 97, no. January, pp. 220-226, 2019, doi: 10.1016/j.engfailanal.2019.01.014.

[10] R. Feng, S. Li, Z. Li, and L. Tian, "Variations of microstructure and properties of 690MPa grade low carbon bainitic steel after tempering," Mater. Sci. Eng. A, vol. 558, pp. 205-210, 2012, doi: 10.1016/j.msea.2012.07.113.

[11] B. Basu, S. M. Tripathi, and V. V. Modak, "Thermomechanically-controlled processing for producing ship-building steels," Def. Sci. J., vol. 55, no. 1, pp. 91-101, 2005.

[12] E. J. Czyryca, "Development of Low-Carbon , Copper-Strengthened HSLA Steel Plate for Naval Ship Construction."

[13] A. Sharma, A. Kumar, and R. Tyagi, "Erosive wear analysis of medium carbon dual phase steel under dry ambient condition," *Wear*, vol. 334-335, no. July 2015, pp. 91-98, 2015, doi: 10.1016/j.wear.2014.12.005.

[14] S. Kumar, "Isothermal Transformation Behavior and Microstructural Evolution of Micro-Alloyed Steel," in *Engineering Steels and High Entropy-Alloys*, A. Sharma, S. Kumar, and Z. Duriagina, Eds. IntechOpen Publication Dondon, 2020, pp. 27-36.

[15] L. Ma, J. Han, J. Shen, and S. Hu, "Effects of Microalloying and Heat-Treatment Temperature on the Toughness of 26Cr–3.5Mo Super Ferritic Stainless Steels," Acta Metall. Sin. (English Lett., vol. 27, no. 3, pp. 407-415, 2014, doi: 10.1007/s40195-014-0070-2.

[16] J. Dille *et al.*, "Influence of Heat Treatments on Microstructure and Magnetic Domains in Duplex Stainless

Steel S31803," Metall. Mater. Trans. A Phys. Metall. Mater. Sci., vol. 49, no. 8, pp. 3515-3524, 2018, doi: 10.1007/s11661-018-4721-1.

[17] A. Lambert-Perlade, A. F. Gourgues, and A. Pineau, "Austenite to bainite phase transformation in the heat-affected zone of a high strength low alloy steel," Acta Materialia, vol. 52, no. 8. pp. 2337-2348, 2004, doi: 10.1016/j.actamat.2004.01.025.

[18] A. F. M. Pérez, M. Breda, I. Calliari, G. Y. P. Medina, and R. Sandström, "Detrimental Cr-rich phases precipitation on SAF 2205 duplex stainless steels welds after heat treatment," Soldag. Inspeção, vol. 21, no. 2, pp. 165-171, 2016, doi: 10.1590/0104-9224/SI2102.06.

[19] Y. Mouadji, M. A. Bradai, R. Younes, A. Sad-eddine, and A. Benabbas, "Influence of heat treatment on microstructure and tribological properties of flame spraying Fe-Ni-Al alloy coating," J. Cent. South Univ., vol. 25, no. 3, pp. 473-481, 2018, doi: 10.1007/s11771-018-3751-6.

[20] S. Kumar and S. K. Nath, "Effect of weld thermal cycles on microstructures and mechanical properties in simulated heat affected zone of a HY 85 steel," Trans. Indian Inst. Met., pp. 1-12, 2016, doi: 10.1007/s12666-016-0880-1.

[21] V. Segal, "Review: Modes and processes of severe plastic deformation (SPD)," *Materials (Basel).*, vol. 11, no. 7, 2018, doi: 10.3390/ma11071175.

[22] Y. Estrin and A. Vinogradov, "Extreme grain refinement by severe plastic deformation: A wealth of challenging science," Acta Mater., vol. 61, no. 3, pp. 782-817, 2013, doi: 10.1016/j.actamat.2012.10.038.

[23] S. Kim, Y. Kang, and C. Lee, "Effect of thermal and thermo-mechanical cycling on the boron segregation behavior in the coarse-grained heat-affected zone of low-alloy steel," Mater. Charact., vol. 116, pp. 65-75, 2016, doi: 10.1016/j.matchar.2016.04.004.

[24] F. Njock Bayock, P. Kah, P. Layus, and V. Karkhin, "Numerical and Experimental Investigation of the Heat Input Effect on the Mechanical Properties and Microstructure of Dissimilar Weld Joints of 690-MPa QT and TMCP Steel," *Metals (Basel).*, vol. 9, no. 3, p. 355, 2019, doi: 10.3390/met9030355.

[25] S. K. Rajput, G. P. Chaudhari, and S. K. Nath, "Characterization of hot deformation behavior of a low carbon steel using processing maps, constitutive equations and Zener-Hollomon parameter," J. Mater. Process. Technol., vol. 237, pp. 113-125, 2016, doi: 10.1016/j.jmatprotec.2016.06.008.

[26] N. Kumar, S. Kumar, S. K. Rajput, and S. K. Nath, "Modelling of Flow Stress and Prediction of Workability by Processing Map for Hot Compression of 43CrNi Steel," ISIJ Int., vol. 57, no. 3, pp. 497-505, 2017, doi: 10.2355/isijinternational.ISIJINT-2016-306.

[27] S. Kumar, S. K. Rajput, N. Kumar, and N. S. K., "Understanding Hot Workability and Flow Stress Prediction through Processing Map with Microstructural Correlation for HY85 Steel," Mater. Perform. Charact., vol. 9, no. 12, pp. 134-149, 2019.

[28] G. E. Dieter, *Mechanical Metallurgy*, Third edit. McGraw-Hill, New York, 1988.

[29] E. Bagherpour, N. Pardis, M. Reihanian, and R. Ebrahimi, "An overview on severe plastic deformation: research status, techniques classification, microstructure evolution, and applications," Int. J. Adv. Manuf. Technol., vol. 100, no. 5-8, pp.

1647-1694, 2019, doi: 10.1007/s00170-018-2652-z.

[30] K. Edalati, "Review on Recent Advancements in Severe Plastic Deformation of Oxides by High-Pressure Torsion (HPT)," Adv. Eng. Mater., vol. 21, no. 1, pp. 1-10, 2019, doi: 10.1002/adem.201800272.

[31] J. Zhang, H. Di, K. Mao, X. Wang, Z. Han, and T. Ma, "Processing maps for hot deformation of a high-Mn TWIP steel: A comparative study of various criteria based on dynamic materials model," Mater. Sci. Eng. A, vol. 587, pp. 110-122, 2013, doi: 10.1016/j.msea.2013.08.036.

[32] S. K. Rajput, G. P. Chaudhari, and S. K. Nath, "Physical Simulation of Hot Deformation of Low-Carbon Ti-Nb Microalloyed Steel and Microstructural Studies," J. Mater. Eng. Perform., vol. 23, no. 8, pp. 2930-2943, 2014, doi: 10.1007/s11665-014-1059-8.

[33] G. Faraji and H. S. Kim, "Review of principles and methods of severe plastic deformation for producing ultrafine-grained tubes," Mater. Sci. Technol. (United Kingdom), vol. 33, no. 8, pp. 905-923, 2017, doi: 10.1080/02670836.2016.1215064.

[34] T. Sakai, A. Belyakov, R. Kaibyshev, H. Miura, and J. J. Jonas, "Dynamic and post-dynamic recrystallization under hot, cold and severe plastic deformation conditions," Prog. Mater. Sci., vol. 60, no. 1, pp. 130-207, 2014, doi: 10.1016/j.pmatsci.2013.09.002.

[35] C. C. F. Kwan and Z. Wang, "The cyclic deformation behavior of severe plastic deformation (SPD) metals and the influential factors," Metals (Basel)., vol. 2, no. 1, pp. 41-55, 2012, doi: 10.3390/met2010041.

[36] D. Samantaray, S. Mandal, C. Phaniraj, and A. K. Bhaduri, "Flow behavior and microstructural evolution during hot deformation of AISI Type 316 L(N) austenitic stainless steel," Mater. Sci. Eng. A, vol. 528, no. 29-30, pp. 8565-8572, 2011, doi: 10.1016/j.msea.2011.08.012.

[37] N. Kumar, S. Kumar, S. K. Rajput, and S. K. Nath, "Modelling of flow stress and prediction of workability by processing map for hot compression of 43CrNi steel," *ISIJ Int.*, vol. 57, no. 3, 2016.

[38] D. Samantaray, A. Chaudhuri, U. Borah, A. K. Bhaduri, and P. Dutta, "Role of grain boundary ferrite layer in dynamic recrystallization of semi-solid processed type 304L austenitic stainless steel," Mater. Lett., vol. 179, pp. 65-68, 2016, doi: 10.1016/j.matlet.2016.05.049.

[39] Z. Yang, F. Zhang, C. Zheng, M. Zhang, B. Lv, and L. Qu, "Study on hot deformation behaviour and processing maps of low carbon bainitic steel," Mater. Des., vol. 66, pp. 258-266, 2015, doi: 10.1016/j.matdes.2014.10.068.

[40] Y. C. Lin, M. S. Chen, and J. Zhang, "Modeling of flow stress of 42CrMo steel under hot compression," Mater. Sci. Eng. A, vol. 499, no. 1-2, pp. 88-92, 2009, doi: 10.1016/j.msea.2007.11.119.

[41] M. Furukawa, Z. Horita, and T. Langdon, *Severe plastic deformation*. 2004.

[42] S. Nemat-Nasser, "Introduction to High Strain Rate Testing," ASM Handb. Int., vol. 8, pp. 427-428, 2000.

[43] V. Raghavan, *Physical metallurgy*, vol. 2. 2012.

[44] Y. C. Lin, M. S. Chen, and J. Zhong, "Prediction of 42CrMo steel flow stress at high temperature and strain rate," Mech. Res. Commun., vol. 35, no. 3, pp. 142-150, 2008, doi: 10.1016/j.mechrescom.2007.10.002.

[45] R. C. Souza, E. S. Silva, A. M. Jorge, J. M. Cabrera, and O. Balancin, “Dynamic recovery and dynamic recrystallization competition on a Nb- and N-bearing austenitic stainless steel biomaterial: Influence of strain rate and temperature,” Mater. Sci. Eng. A, vol. 582, pp. 96-107, 2013, doi: 10.1016/j.msea.2013.06.037.

[46] E. J. Giordani, A. M. Jorge, and O. Balancin, “Proportion of recovery and recrystallization during interpass times at high temperatures on a Nb- and N-bearing austenitic stainless steel biomaterial,” Scr. Mater., vol. 55, no. 8, pp. 743-746, 2006, doi: 10.1016/j.scriptamat.2006.05.015.

[47] A. Heidarzadeh *et al.*, “Friction stir welding/processing of metals and alloys: A comprehensive review on microstructural evolution,” *Prog. Mater. Sci.*, no. October, p. 100752, 2020, doi: 10.1016/j.pmatsci.2020.100752.

[48] B. Mirzakhani, M. T. Salehi, S. Khoddam, S. H. Seyedein, and M. R. Aboutalebi, “Investigation of dynamic and static recrystallization behavior during thermomechanical processing in a API-X70 microalloyed steel,” J. Mater. Eng. Perform., vol. 18, no. 8, pp. 1029-1034, 2009, doi: 10.1007/s11665-008-9338-x.

[49] F. Montheillet, J. Lépinoux, D. Weygand, and E. Rauch, “Dynamic and static recrystallization,” Adv. Eng. Mater., vol. 3, no. 8, pp. 587-589, 2001, doi: 10.1002/1527-2648(200108)3:8<587::AID-ADEM587>3.0.CO;2-V.

[50] N. Hansen and C. Y. Barlow, *Plastic Deformation of Metals and Alloys*, Fifth Edit., vol. 1. Elsevier, 2014.

[51] N. Dudova, A. Belyakov, T. Sakai, and R. Kaibyshev, “Dynamic recrystallization mechanisms operating in a Ni-20%Cr alloy under hot-to-warm working,” Acta Mater., vol. 58, no. 10, pp. 3624-3632, 2010, doi: 10.1016/j.actamat.2010.02.032.

[52] A. Belyakov, T. Sakai, H. Miura, and R. Kaibyshev, “Grain Refinement under Multiple Warm Deformation in 304 type Austenitic Stainless Steel,” ISIJ Int., vol. 39, no. 6, pp. 592-599, 1999.

[53] M. C. Poletti, R. Buzolin, S. Kumar, P. Wang, and T. F. J. Simonet-Fotso, “Microstructure evolution of ti-5al-5v-5mo-3cr after hot deformation at large and moderate strains,” *Mater. Sci. Forum*, vol. 941 MSF, pp. 1443-1449, 2018, doi: 10.4028/www.scientific.net/MSF.941.1443.

[54] P. R. Rios, F. Siciliano, H. R. Z. Sandim, R. L. Plaut, and A. F. Padilha, “Nucleation and growth during recrystallization,” Mater. Res., vol. 8, no. 3, pp. 225-238, 2005, doi: 10.1590/S1516-14392005000300002.

[55] M. Tikhonova, R. Kaibyshev, and A. Belyakov, “Microstructure and Mechanical Properties of Austenitic Stainless Steels after Dynamic and Post-Dynamic Recrystallization Treatment,” Adv. Eng. Mater., vol. 20, no. 7, pp. 1-27, 2018, doi: 10.1002/adem.201700960.

[56] B. Gong, X. W. Duan, J. S. Liu, and J. J. Liu, “A physically based constitutive model of As-forged 34CrNiMo6 steel and processing maps for hot working,” *Vacuum*, vol. 155, no. April, pp. 345-357, 2018, doi: 10.1016/j.vacuum.2018.06.022.

[57] D. Samantaray, S. Mandal, V. Kumar, S. K. Albert, A. K. Bhaduri, and T. Jayakumar, “Optimization of processing parameters based on high temperature flow behavior and microstructural evolution of a nitrogen enhanced 316L(N) stainless steel,” *Mater. Sci. Eng. A*, vol. 552, no. July 2016, pp. 236-244, 2012, doi: 10.1016/j.msea.2012.05.036.

[58] T. Xi, C. Yang, M. Babar Shahzad, and K. Yang, “Study of the processing

map and hot deformation behavior of a Cu-bearing 317LN austenitic stainless steel," Mater. Des., vol. 87, pp. 303-312, 2015, doi: 10.1016/j.matdes.2015.08.011.

[59] H. Mecking and U. F. Kocks, "Kinetics of flow and strain-hardening," *Acta Metall.*, vol. 29, no. 11, pp. 1865-1875, 1981, doi: https://doi.org/10.1016/0001-6160(81)90112-7.

[60] N. D. Ryan and H. J. McQueen, "Flow stress, dynamic restoration, strain hardening and ductility in hot working of 316 steel," *J. Mater. Process. Technol.*, vol. 21, no. 2, pp. 177-199, 1990, doi: https://doi.org/10.1016/0924-0136(90)90005-F.

[61] H. J. McQueen, S. Yue, N. D. Ryan, and E. Fry, "Hot working characteristics of steels in austenitic state," *J. Mater. Process. Technol.*, vol. 53, no. 1, pp. 293-310, 1995, doi: https://doi.org/10.1016/0924-0136(95)01987-P.

[62] P. Gao, M. Fu, M. Zhan, Z. Lei, and Y. Li, "Deformation behavior and microstructure evolution of titanium alloys with lamellar microstructure in hot working process: A review," J. Mater. Sci. Technol., vol. 39, pp. 56-73, 2020, doi: 10.1016/j.jmst.2019.07.052.

[63] S. Patra, V. Kumar, A. Haldar, and D. Chakrabarti, "Effect of hot-deformation on micro-texture in ultra-fine grained HSLA steel," vol. 703, no. January, pp. 439-442, 2012, doi: 10.4028/www.scientific.net/MSF.702-703.439.

[64] F. Ren, F. Chen, J. Chen, and X. Tang, "Hot deformation behavior and processing maps of AISI 420 martensitic stainless steel," J. Manuf. Process., vol. 31, pp. 640-649, 2018, doi: 10.1016/j.jmapro.2017.12.015.

[65] B. Srinivas, C. Srinivasu, B. Mahesh, and M. Aqheel, "A Review on Severe Plastic Deformation," *Int. J. Adv. Mater. Manuf. Charact.*, vol. 3, no. 1, pp. 291-295, 2013, doi: 10.11127/ijammc.2013.02.053.

[66] M. K. Mishra, A. G. Rao, R. Sarkar, B. P. Kashyap, and N. Prabhu, "Effect of Preaging Deformation on Aging Characteristics of 2507 Super Duplex Stainless Steel," J. Mater. Eng. Perform., vol. 25, no. 2, pp. 374-381, 2016, doi: 10.1007/s11665-015-1840-3.

[67] P. O. Malta, F. L. Dias, A. C. M. de Souza, and D. B. Santos, "Microstructure and texture evolution of duplex stainless steels with different molybdenum contents," *Mater. Charact.*, vol. 142, no. June, pp. 406-421, 2018, doi: 10.1016/j.matchar.2018.06.006.

[68] J. L. del Abra-Arzola *et al.*, "Study of the effect of sigma phase precipitation on the sliding wear and corrosion behaviour of duplex stainless steel AISI 2205," *Wear*, vol. 400-401, no. August 2017, pp. 43-51, 2018, doi: 10.1016/j.wear.2017.12.019.

[69] R. Mahnken, A. Schneidt, T. Antretter, U. Ehlenbröker, and M. Wolff, "Multi-scale modeling of bainitic phase transformation in multi-variant polycrystalline low alloy steels," Int. J. Solids Struct., vol. 54, pp. 156-171, 2015, doi: 10.1016/j.ijsolstr.2014.10.021.

[70] M. F. Ashby, "The deformation of plastically non-homogeneous materials," Philos. Mag., vol. 21, no. 170, pp. 399-424, 1970, doi: 10.1080/14786437008238426.

[71] H. J. McQueen, "Dynamic Recovery and Recrystallization," Encycl. Mater. Sci. Technol., pp. 2375-2381, 2001, doi: 10.1016/b0-08-043152-6/00419-8.

[72] K. P. Kolhe and C. K. Datta, "Prediction of microstructure and mechanical properties of multipass SAW," J. Mater. Process. Technol., vol. 197, no. 1-3, pp. 241-249, 2008, doi: 10.1016/j.jmatprotec.2007.06.066.

[73] J. Y. Choi, S. W. Hwang, and K. T. Park, "Twinning-induced plasticity aided high ductile duplex stainless steel," Metall. Mater. Trans. A Phys. Metall. Mater. Sci., vol. 44, no. 2, pp. 597-601, 2013, doi: 10.1007/s11661-012-1579-5.

[74] H. Gholizadeh, "The Influence of Alloying and Temperature on the Stacking-fault Energy of Iron-based Alloys," *PhD Diss. Mont. Leoben*, no. May, p. 193 pp., 2013.

[75] B. G. Prusty and A. Banerjee, "Structure-property correlation and constitutive description of structural steels during hot working and strain rate deformation," *Materials (Basel).*, vol. 13, no. 3, 2020, doi: 10.3390/ma13030556.

[76] F. Tehovnik, B. Arzenšek, B. Arh, D. Skobir, B. Pirnar, and B. Žužek, "Microstructure evolution in SAF 2507 super duplex stainless steel," Mater. Tehnol., vol. 45, no. 4, pp. 339-345, 2011.

[77] H. Li *et al.*, "Deformation Characteristic and Constitutive Modeling of 2707 Hyper Duplex Stainless Steel under Hot Compression," *Metals (Basel).*, vol. 6, no. 9, p. 223, 2016, doi: 10.3390/met6090223.

[78] D. Samantaray, V. Kumar, a K. Bhaduri, and P. Dutta, "Microstructural Evolution and Mechanical Properties of Type 304 L Stainless Steel Processed in Semi-Solid State," Int. J. Metall. Eng., vol. 2, no. 2, pp. 149-153, 2013, doi: 10.5923/j.ijmee.20130202.06.

[79] J. Humphreys, G. S. Rohrer, and A. Rollett, *Recrystallization and Related Annealing Phenomena: Second Edition*. 2017.

[80] C. P. Wang, F. G. Li, W. Lei, and H. J. Qiao, "Review on modified and novel techniques of severe plastic deformation," Sci. China Technol. Sci., vol. 55, no. 9, pp. 2377-2390, 2012, doi: 10.1007/s11431-012-4954-y.

[81] R. Jacob, S. Raman Sankaranarayanan, and S. P. Kumaresh Babu, "Recent advancements in manganese steels-A review," Mater. Today Proc., vol. 27, pp. 2852-2858, 2019, doi: 10.1016/j.matpr.2020.01.296.

[82] M. Sabzi and M. Farzam, "Hadfield manganese austenitic steel: A review of manufacturing processes and properties," *Mater. Res. Express*, vol. 6, no. 10, 2019, doi: 10.1088/2053-1591/ab3ee3.

[83] R. Hadfield, "HADFIELD'S MANGANESE STEEL.," Science, vol. 12, no. 306, pp. 284-286, Dec. 1888, doi: 10.1126/science.ns-12.306.284-a.

[84] L. Rémy and A. Pineau, "Twinning and strain-induced f.c.c. → h.c.p. transformation on the mechanical properties of CoNiCrMo alloys," Mater. Sci. Eng., vol. 26, no. 1, pp. 123-132, 1976, doi: 10.1016/0025-5416(76)90234-2.

[85] O. Bouaziz, H. Zurob, B. Chehab, J. D. Embury, S. Allain, and M. Huang, "Effect of chemical composition on work hardening of Fe-Mn-C TWIP steels," Mater. Sci. Technol., vol. 27, no. 3, pp. 707-709, 2011, doi: 10.1179/026708309X12535382371852.

[86] S. Allain, J. P. Chateau, O. Bouaziz, S. Migot, and N. Guelton, "Correlations between the calculated stacking fault energy and the plasticity mechanisms in Fe-Mn-C alloys," *Mater. Sci. Eng. A*, vol. 387-389, no. 1-2 SPEC. ISS., pp. 158-162, 2004, doi: 10.1016/j.msea.2004.01.059.

[87] Z. C. Yanushkevich, D. A. Molodov, A. N. Belyakov, and R. O. Kaibyshev, "Recrystallization kinetics of an austenitic high-manganese steel subjected to severe plastic deformation," Russ. Metall., vol. 2016, no. 9, pp. 812-819, 2016, doi: 10.1134/S0036029516090184.

[88] W. Yan, L. Fang, K. Sun, and Y. Xu, "Effect of surface work hardening on wear behavior of Hadfield steel," *Mater. Sci. Eng. A*, vol. 460-461, pp. 542-549, 2007, doi: https://doi.org/10.1016/j.msea.2007.02.094.

[89] M. Soleimani, A. Kalhor, and H. Mirzadeh, "Transformation-induced plasticity (TRIP) in advanced steels: A review," *Mater. Sci. Eng. A*, vol. 795, no. August, 2020, doi: 10.1016/j.msea.2020.140023.

[90] S. Sevsek, C. Haase, and W. Bleck, "Strain-rate-dependent deformation behavior and mechanical properties of a multi-phase medium-manganese steel," *Metals (Basel).*, vol. 9, no. 3, 2019, doi: 10.3390/met9030344.

[91] M. Bahramyan, R. T. Mousavian, and D. Brabazon, "Study of the plastic deformation mechanism of TRIP-TWIP high entropy alloys at the atomic level," *Int. J. Plast.*, vol. 127, no. June 2019, p. 102649, 2020, doi: 10.1016/j.ijplas.2019.102649.

[92] F. Javadzadeh Kalahroudi, H. Koohdar, H. R. Jafarian, Y. Haung, T. G. Langdon, and M. Nili-Ahmadabadi, "On the microstructure and mechanical properties of an Fe-10Ni-7Mn martensitic steel processed by high-pressure torsion," *Mater. Sci. Eng. A*, vol. 749, no. January, pp. 27-34, 2019, doi: 10.1016/j.msea.2019.02.002.

[93] H. Fu *et al.*, "Dynamic behaviors and microstructure evolution of iron–nickel based ultra-high strength steel by SHPB testing," *Metals (Basel).*, vol. 10, no. 1, 2020, doi: 10.3390/met10010062.

[94] L. S. Toth *et al.*, "Modeling the effect of primary and secondary twinning on texture evolution during severe plastic deformation of a twinning-induced plasticity steel," *Materials (Basel).*, vol. 11, no. 5, 2018, doi: 10.3390/ma11050863.

[95] M. C. Somani and L. P. Karjalainen, "Innovative Approaches in Physical Simulation and Modeling for Optimal Design and Processing of Advanced High Strength Steels," Mater. Manuf. Process., vol. 25, no. 1-3, pp. 133-141, 2010, doi: 10.1080/10426910903158223.

[96] J. H. Ryu, P. H. K. D. H. Bhadeshia, and P. D.-W. Suh, "Hydrogen Embrittlement in TRIP and TWIP Steels," *Grad. Inst. Ferr. Technol.*, vol. Ph.D., 2012.

[97] G. Frommeyer, U. Brüx, and P. Neumann, "Supra-ductile and high-strength manganese-TRIP/TWIP steels for high energy absorption purposes," ISIJ Int., vol. 43, no. 3, pp. 438-446, 2003, doi: 10.2355/isijinternational.43.438.

[98] J. Y. Kang, J. G. Kim, S. K. Kim, K. G. Chin, S. Lee, and H. S. Kim, "Outstanding mechanical properties of high-pressure torsion processed multiscale TWIP-cored three layer steel sheet," Scr. Mater., vol. 123, pp. 122-125, 2016, doi: 10.1016/j.scriptamat.2016.06.009.

Chapter 2

Explosive Welding Process to Clad Materials with Dissimilar Metallurgical Properties

Bir Bahadur Sherpa and Reetu Rani

Abstract

Explosive welding is a solid-state process, which is an advanced form of joining two metal plates with dissimilar metallurgical properties, irrespective of the differences in physical and chemical properties. In this process, high pressure of explosive is used to accelerate one metal plate over another to form the bimetallic product. The pressure needs to be sufficiently high and for enough length of time to achieve inter-atomic bonds. During the explosive welding process, a jetting phenomenon occurs at the collision point which cleans the top oxide layer over metals and leaves the virgin surfaces that help in the joining process. The metals are joined without losing their pre-bonded properties with higher bond strengths than the strength of the weaker parent material. There are various critical factors such as explosive type, mass of explosive, stand-off distance, type of plate material, velocity of detonation etc. which affect the bond quality. Researchers mainly play with all these parameters to bring out the best characteristics of the bimetallic product that can be used for the desired applications such as heat exchanger, pressure vessels etc.

Keywords: bond strength, dissimilar materials, detonation velocity, explosive welding, metallurgical properties

1. Introduction

Welding is a process of joining two materials together through pressure, heat and sometimes with the addition of filler materials. The important condition for any welding technique is that the two surfaces that need to be joined should be cleaned and uncontaminated. Moreover, if the two surfaces are brought together in such a way that the surfaces exchange the outer orbit of the valence electrons and form interatomic bonds, the weld formed will be very strong in terms of mechanical properties. But this kind of bond formation is not possible through conventional means. In most of the welding techniques melting is involved in joining the two components. There are also some welding processes such as solid-state welding processes where heat required is below the melting point of the base material being welded and therefore, no melting is observed during joining for example ultrasonic welding [1, 2], friction welding [3–5], cold welding [6], explosive welding [7–10] and diffusion welding [11, 12]. All of the welding methods have some advantages and disadvantages in their particular field and are applied as per the need of the applications. In the current world, there is an increasing trend of using dissimilar material combinations for various applications such as automobile, shipbuilding, military, aerospace and oil industries etc. The bi-metallic product takes the mechanical advantage of both the materials such as wear resistance, corrosion resistance, high tensile strength and lightweight. To meet such requirements many researchers are extensively working in this field to produce such combinations. In which explosive welding is considered as one of the potential welding technique and is gaining more attention due to its vast features as mentioned [13, 14]. Explosive welding is one of the solid-state welding processes in which explosive energy is used to create a high-velocity impact collision between the two plates to be joined. The process can join a wide area of non-compatible material combination irrespective of the difference in mechanical and chemical properties and which cannot be joined by any other conventional means. It is a surface bond welding, which provides a strong metallurgical bond at the molecular level and provides strength higher than the base materials [15]. There are various applications of explosive welding products such as in cryogenic pressure vessels [16], scram jet engine components [17], shipbuilding application [18].

2. Working principle of explosive welding

In the explosive welding process, the explosive is used as a source of energy to accelerate one of the metal plates into another. **Figure 1a** shows the initial set-up of the explosive welding process showing the two plates i.e. base plate which is kept stationary and the movable flyer plate is kept at a particular calculated distance called stand-off distance. The explosive box is placed with a buffer sheet over the plates. This buffer sheet protects the flyer plate from damage due to explosion. To initiate the main explosive detonator is used, which is placed above explosive. **Figure 1b** shows the schematic diagram after the detonation of explosive has initiated in the explosive welding process. Here we can observe the collision point, where the two plates collide and the bond formation occurs. Along with this jetting phenomenon is witnessed which is one of the most important criteria and also an essential condition for bond formation. Jetting occurs during an oblique collision at the collision point, in which it cleans the mating surfaces and Leaves behind a virgin surface free from oxide layers and contaminants. This helps to interact two mating materials at the atomic level when subjected to high impact pressure waves arising from the explosion effects. This process is capable of joining large surface area due to its ability to distribute high energy density. Explosive welding can be basically

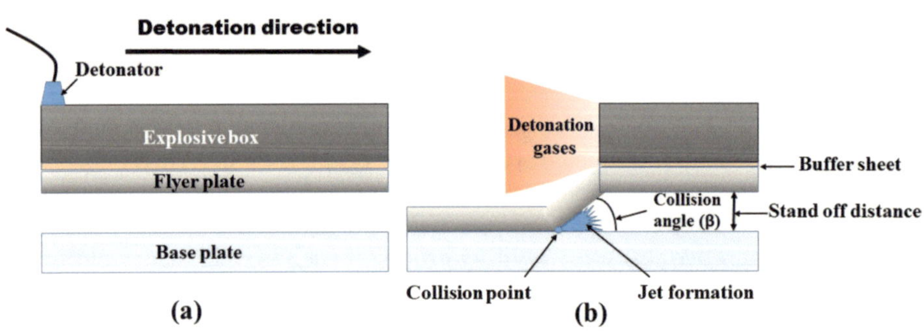

Figure 1.
Schematic diagram of explosive welding process in parallel set-up, a) initial set-up, b) after the explosion has initiated.

defined in two steps; first jet phenomenon occurs and cleans up the oxide layers and second, the high impact pressure forces the mating surfaces into such intimate contact that they meet at the interatomic level and results in strong metallurgical bond.

2.1 Plastic deformation in explosive welding process

In explosive welding process due to detonation effect of explosive many critical phenomena occur such as release of large gas product i.e. explosion, high impact collision between mating surfaces, high temperature, generation of heat, plastic deformation in the metal plates, pressure generation, jetting and bonding occurs for a very short period of time i.e. microseconds [19–21]. Out of these, plastic deformation that occurs at the weld interface due to high impact pressure is considered as one of the important factor responsible for good bond formation. Pastic deformation in explosive welding process occurs when pressure at the collision front overcomes the yield strength of the materials. Through plastic deformation an intimate contact is formed where the two mating surfaces are brought too close together that atomic reaction occurs between the mating surfaces [22–24]. Plastic deformation can be examined using visioplastic methods without disturbing the original properties of materials. The most distinctive form of plastic deformation is the wave formation in explosive welding [25]. Occurrence of high plastic deformation of the mating surfaces lead to grain refinement [26]. Difference in grain size adjacent to weld interface is observed due to severe plastic deformation [27]. Various researchers have witnessed high hardness value at the weld interface of explosively welded specimens in microhardness examination study. It was mainly attributed to intense plastic deformation developed across the weld interface. The level of plastic deformation in explosively welded specimens decrease gradually with increase in distance from the weld interface [28–30].

2.2 Types of experimental set-up

There are two types of explosive welding set-up i.e. parallel and the inclined set-up [31]. **Figure 1** shows the parallel set-up where the two plates are placed parallel to each other. This kind of configuration is used for joining large and thick plates. While the inclined set-up is shown in **Figure 2** in which flyer plates are inclined at a particular angle (α). This kind of configuration is generally applied for joining small and thin plates.

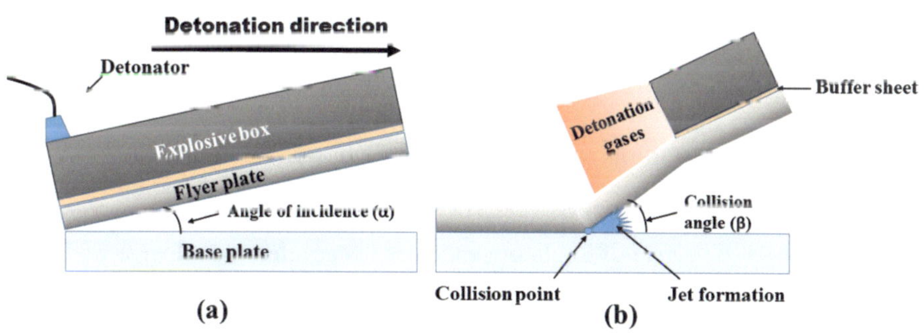

Figure 2.
Schematic diagram of explosive welding process in inclined set-up a) initial set-up, b) after the explosion has initiated.

2.3 Terminology used in explosive welding

Base plate: It is the one which is placed at the open ground or at the anvil. This is kept stationary and is the one on which the cladding is performed. Both the base plate and the flyer plate are cleaned thoroughly and polished gently before welding.

Flyer plate: It is the one which is placed above the base plate and during collision this plate hits onto the base plate. The selection of flyer plate and base plate is done on the basis of mass per unit area, whoever is less is placed as flyer plate. As compared to base plate it has the lowest density as well as tensile strength.

Stand-off distance: It is the one which maintains the distance between the flyer plate and the base plate. Stand-off distance helps the flyer plate to accelerate and acquire the required impact velocity to generate jetting. Apart from this, it also provides the exit path to the jet and the air formed between the flyer and base plate during the collision. In general stand-off distance is kept half or equal to the thickness of flyer plate.

Buffer sheet: This sheet is placed over the flyer plate. It is made up of rubber or PVC. The main role of this sheet is to protect the flyer plate from damage which can occur during collision due to explosion effects.

Explosive box: It is placed over the metal plates to be welded. This acts as a source of energy which provides the required forces to weld the materials. Explosive can be used as powder, slurry or sheet form which is spread over the buffer sheet uniformly.

Detonator: This is placed at the top of the explosive box. The main function of the detonator is to help in initiating the main explosive. The detonator is detonated with the help of dynamo placed at some distance from the trial site.

3. Different parameters affecting the explosive welded products

There are various parameters which influence the final product of the explosive welding process. Therefore, careful control of welding parameters is very critical. The criteria for selection of the welding parameters depends upon the mechanical properties of the matting surfaces [32, 33]. Many researchers change the magnitude of these parameters by playing with the different parameters such as detonation velocity, stand-off distance, explosive type etc. The various process parameters are discussed below.

Explosive: In explosive welding, controlled energy of explosive is used to accelerate the flyer plate and help to impact on to the base plate, to produce a strong metallurgical bond. Explosive is generally characterized by their velocity of detonation (VoD) and density. In most of the engineering materials, the velocity of sound is between 4.5–6 km/s and most of the common explosives have VoD ranging between 6 and 7 km/s. Therefore, high VoD in explosive welding is not preferable as in case of joining the weld will get dismantle or in some cases it will destroy the material. In explosive welding, VoD is mostly applied in the range of 2–3 km/s to obtain a uniform detonation across the joining metal plates [32–34]. Many researchers have worked with different explosives to obtain a sound weld. A. Loureiro et al. have studied the effect of explosive mixture i.e. emulsion explosive with two different sensitizers i.e. hollow glass microspheres (HGMS) and expanded polystyrene spheres (EPS) on the weld interface of copper-aluminum. They observed improved surface using HGMS and higher wave amplitude was witnessed by employing EPS [35]. Similarly, many works related to explosive optimization have been done in the past in explosive welding [36, 37]. Recently Sherpa et al. have developed a low velocity of detonation (VoD) explosive welding process (LVEW) in which VoD was less than 2 km/s and obtained a sound joint [38]. Some of the explosives used for explosive welding process are shown in **Table 1**.

Stand-off distance: Stand-off distance is normally selected based on the thickness of the flyer plate and the explosive parameters. It is one of the critical parameters which influence the bond quality. Stand-off distance is selected basically to provide necessary dynamic bend angle and the impact velocity for proper bond to form. Durgutlu et al. studied the effect of stand-off distance on copper and stainless steel bond. They observed an increase in wavelength and amplitude of the wave with an increase in stand-off distance. As well as hardness value across the weld interface also increased with increasing stand-off distance [48]. M.R. Jandaghi et al. studied the effect of stand-off distance on the copper and aluminum interface. They observed that with an increase in stand-off distance, plastic deformation, kinetic energy at the collision point and as well as the melting increases at the weld interface which lead to the increase in corrosion rate [49].

Flyer plate velocity: It is the velocity at which the flyer plate strike into the base plate after the detonation has started. To obtain good bonding, the flyer plate velocity should be in the described limits i.e. between the minimum and maximum flyer velocity. Experimenting with flyer plate velocity above defined range can lead to certain defects such as melting zone, cracks, brittle phases, bend and damage of flyer plate [50, 51].

Collision angle (β): It is the angle formed between the flyer plate and the base plate during the collision process. Collision angle should be selected very carefully to meet the requirement of the bonding parameters. If the angle is selected below the critical collision angle, a jet-less phenomenon will occur and if β is chosen above defined limit it will cause entrapment of jet [33, 52].

Collision velocity (Vc): It is the velocity with which collision point moves along the area being welded. For proper welding to occur there should be some plastic flow ahead of the collision point. Hence, the collision point velocity should be less than the sonic velocity in the metals. The smooth interface is observed at lower collision velocity while the wavy interface is observed at higher collision velocity at the weld interface. Increasing the collision velocity may also increase the chances of melt pockets across the weld interface [20].

3.1 Weldability window

The condition that should satisfy for proper bonding to take place is defined by weldability window. Detailed view with the description of weldability window is shown in **Figure 3**. It is plotted between collision angle (β) and collision velocity (Vp), where it is well defined by four different lines [50, 51]. The first limit is placed at the rightmost side in which formation of the jet at the collision point

Explosive	Velocity of detonation (m/s)	Density (kg/m^3)	Ref.
ANFO (ammonium nitrate with fuel oil)	2300–2800	650–700	[39, 40]
SEP	7000	1300	[41, 42]
Emulsion explosive	2200	1150	[43]
Elbar-5	3000–3800	700–800	[44–46]
PAVEX	2000–3000	530	[37, 47]

Table 1.
Different explosives used in the explosive welding process.

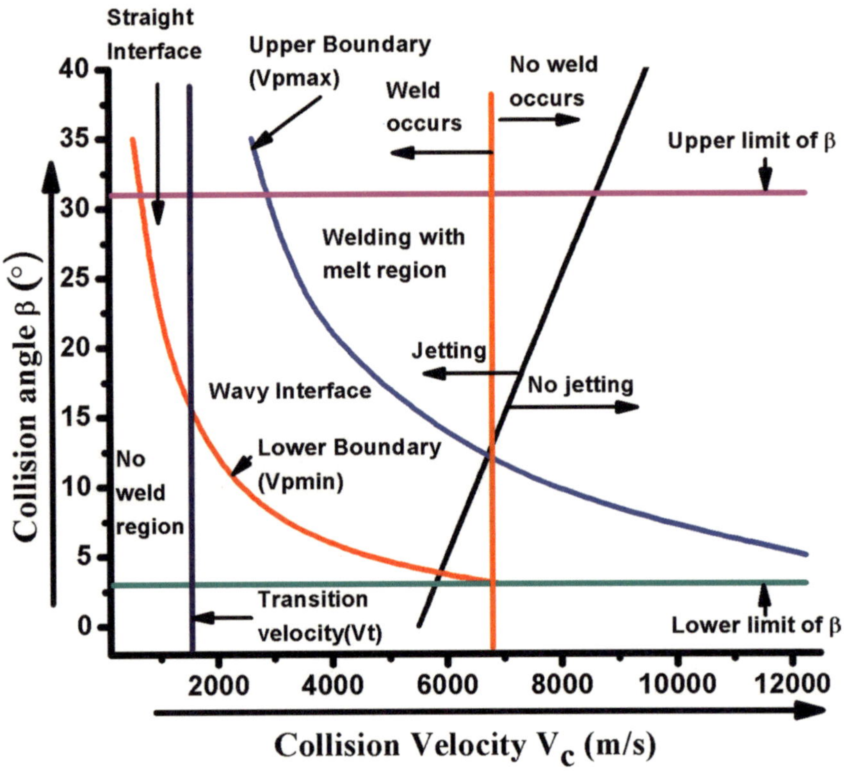

Figure 3.
Weldability window concepts for explosive welding process.

is considered. As jetting is one of the important criteria in explosive welding. Abrahamson [53] linked welding velocity with the collision angle β as shown in Eq. 1 for the first limit. The second limit is placed at the left side of the weldability window which is related to the formation of wavy morphology at the weld interface. Cowan et al. introduced Reynolds number for describing the laminar and turbulent flow [20] as shown in Eq. 2. The third limit is related to the minimum flyer plate velocity (Vpmin) which ensure that the impact pressure developed at the collision point exceeds the yield strength of the materials. Lower boundary equation was developed for third limit as shown in Eq. 3. While the fourth limit corresponds to the maximum flyer plate velocity (Vpmax) which maintains the required impact pressure below the value so that the melting does not occur at the weld interface. To avoid melting Eq. 4 was developed by Wittman [50]. Therefore, in order to obtain good bond, selection of welding parameters should be with in the described limits of weldability window [20, 34, 50, 54].

$$V_c = \frac{\beta}{10} + 5.5 \tag{1}$$

$$R_t = \frac{(\rho_a + \rho_b) V_c^2}{2(H_a + H_b)} \tag{2}$$

Where ρ_a & ρ_b : ρ Density of flyer plate and base plate

H_a & H_b : Hardness value of flyer plate and base plate

R_t : Reynolds number

$$\sin\beta = k\sqrt{\frac{H_V}{\rho V_C^2}} \tag{3}$$

Where H_V : Vickers hardness no.
P: Density of the material
K: Constant value
Take value 0.6: Plate surface is very clean
1.2: Imperfectly cleaned plate surface

$$\sin\left(\frac{\beta}{2}\right) = \frac{1}{2N}\frac{(T_m C_o)^{1/2}}{V_c^2}\left(\frac{KC_p C_b}{\rho h}\right)^{1/4} \tag{4}$$

Where Tm: Melting temperature,
Cp: Specific heat capacity,
K: Thermal conductivity,
h: Thickness of flyer plate,
Cb: Bulk sound speed

3.2 Different materials combination joined by explosive welding

Explosive welding process is capable of joining similar and dissimilar material combinations irrespective of the difference in physical and chemical properties.

Similar materials combinations			
Material combination	**Welding configuration**	**Explosive used**	**Ref.**
Al alloy -Al alloy	Tube	PETN	[55]
Steel-steel plates	Parallel set-up	Elbar-5	[56–58]
Steel-steel	Cylindrical	Emulsion explosive/ ANFO	[36]
Copper-copper	—	—	[59]
Copper-copper alloy	Parallel set-up	Powder emulsion explosive	[43]
Dissimilar materials combinations			
Material combination	**Welding configuration**	**Interlayer used**	**Ref.**
Titanium and magnesium alloy AZ31	Inclined set-up (Under water)	Thin AZ31	[60]
Aluminum to stainless steel	Parallel set-up	Cu, Ti & Ta	[16]
C103 niobium alloy and C263 nimonic alloy	Parallel set-up	Not used	[17]
Titanium and aluminum	Parallel set-up	Not used	[61, 62]
Aluminum and copper	Parallel set-up	Al5052, Cu & SS304	[63]
Sn and Cu	Inclined set-up (Under water)	Not used	[41]
Al and Mg alloy	Parallel set-up	Not used	[64, 65]
Aluminum and carbon steel and Aluminum-stainless steel	Parallel set up	Aluminum AA1050	[66]
Aluminum and copper	Parallel set-up	Not used	[67–69]
Aluminum and steel	Parallel set-up	Not used	[70–73]

Table 2.
Material combinations joined using explosive welding process.

The various material combinations joined using explosive welding process i.e. similar and dissimiliar combinations are given in **Table 2.** In this process, different authors have also used the concept of the interlayer to minimize the kinetic energy loss as well as the formation of meting zone at the weld interface.

3.3 Important points in explosive welding process

Following points should be considered for explosive welding process to produce a strong metallurgical bond.

- The pressure generated at the collision point should be enough in magnitude so as to exceed the dynamic elastic limits of the mating materials in order to ensure that deformation has occurred at the weld interface [74].

- Stand-off distance should be calculated properly to ensure that the flyer plate can accelerate to the required impact velocity needed for good bonding. Use of high stand-off distance will result in edge instability and can also affect the bonding quality [74, 75].

- The explosive used should provide sufficient energy in order to accelerate the flyer plate to the preferred velocity. The high detonation velocity of explosive should also be avoided as it can lead to spalling and damage of the joining materials. Therefore, the velocity of detonation must be less than 120% of the sonic velocity of the materials being welded [76, 77].

- Flyer plate velocity (Vp) and collision velocity (Vc) should be less than the velocity of sound in either of the participant material. In order that the reflected stress waves do not interfere with the incident wave at the collision point [19, 78, 79].

4. Conclusions

a. Explosive welding is a solid-state welding process capable of joining any material combination which cannot be joined by any other conventional means. It can join materials irrespective of the difference in chemical and physical properties.

b. Jetting is one of the important criteria in explosive welding process which removes the oxide layers present at the mating surfaces. This jetting freely exit at the corners of the joint if the welding parameters are selected properly else if it gets trapped will result in the defects.

c. Plastic deformation is caused due to high impact pressure and is considered as one of the important condition for joint formation in explosive welding process. Plastic deformation leads to the intimate contact of the two mating surfaces and results in strong metallurgical bond formation. It is responsible for grain refinement as well as increase in hardness value across the weld interface of explosively welded samples.

d. To obtain a good bond, various welding parameters such as type of explosive, stand-off distance, flyer plate velocity, and collision velocity need to be selected very carefully. As these parameters will directly or indirectly affect the product of the weld.

e. During explosive welding, there are various defects which are uncounted especially intermetallic formation at the weld interface. To minimize these defects researchers are using different approaches such as interlayer concept and low velocity of detonation explosives which will reduce the kinetic energy loss at the collision point.

f. In the explosive welding process, we can join two materials and take the mechanical advantage of both the materials in the final product. Due to its enormous advantages, it has great application in the field of aerospace, automobiles, oil industries, defense and ship industries.

Acknowledgements

I would like to deeply acknowledge Dr. Pal Dinesh Kumar, Scientist-F, Joint director (MEMWD), Terminal Ballistics Research Laboratory (TBRL-DRDO) and Dr. Sachin Tyagi, Sr. Scientist, Central Scientific Instruments Organization (CSIO-CSIR), Chandigarh, India for their motivation and guidance during the whole journey.

Conflict of interest

The authors declare no conflict of interest.

Author details

Bir Bahadur Sherpa[1,2]* and Reetu Rani[1,2]

1 Academy of Scientific and Innovative Research (AcSIR), Ghaziabad-201002, India

2 CSIR-Central Scientific Instruments Organisation (CSIO), Sector-30, Chandigarh-160030, India

*Address all correspondence to: sherpa7419@gmail.com

References

[1] Benatar, A. and T.G. Gutowski, *Ultrasonic welding of peek graphite APC-2 composites.* Polymer Engineering & Science, 1989. **29**(23): p. 1705-1721.

[2] Watanabe, T., H. Sakuyama, and A. Yanagisawa, *Ultrasonic welding between mild steel sheet and Al–Mg alloy sheet.* Journal of Materials Processing Technology, 2009. **209**(15-16): p. 5475-5480.

[3] Meshram, S., T. Mohandas, and G.M. Reddy, *Friction welding of dissimilar pure metals.* Journal of Materials Processing Technology, 2007. **184**(1-3): p. 330-337.

[4] Uday, M., et al., *Advances in friction welding process: a review.* Science and technology of Welding and Joining, 2010. **15**(7): p. 534-558.

[5] Dey, H., et al., *Joining of titanium to 304L stainless steel by friction welding.* Journal of Materials Processing Technology, 2009. **209**(18-19): p. 5862-5870.

[6] Lu, Y., et al., *Cold welding of ultrathin gold nanowires.* Nature nanotechnology, 2010. **5**(3): p. 218-224.

[7] Akbari-Mousavi, S., L. Barrett, and S. Al-Hassani, *Explosive welding of metal plates.* Journal of materials processing technology, 2008. **202**(1-3): p. 224-239.

[8] Mousavi, S.A. and S. Al-Hassani, *Finite element simulation of explosively-driven plate impact with application to explosive welding.* Materials & Design, 2008. **29**(1): p. 1-19.

[9] Wronka, B., *Testing of explosive welding and welded joints: joint mechanism and properties of explosive welded joints.* Journal of materials science, 2010. **45**(15): p. 4078-4083.

[10] Mousavi, A.A. and S. Al-Hassani, *Numerical and experimental studies of the mechanism of the wavy interface formations in explosive/impact welding.* Journal of the Mechanics and Physics of Solids, 2005. **53**(11): p. 2501-2528.

[11] Aydın, K., Y. Kaya, and N. Kahraman, *Experimental study of diffusion welding/bonding of titanium to copper.* Materials & Design, 2012. **37**: p. 356-368.

[12] Arik, H., et al., *Weldability of Al4C3–Al composites via diffusion welding technique.* Materials & design, 2005. **26**(6): p. 555-560.

[13] Blazynski, T.Z., *Explosive welding, forming and compaction*. 2012: Springer Science & Business Media.

[14] Crossland, B., *Development of exlosive welding and its application in engineering.* Metals materials, 1971. **5**(12): p. 402-413.

[15] Sherpa, B.B., et al., *Explosive Welding of Al-MS Plates and its Interface Characterization.* Explosion Shock Waves and High Strain Rate Phenomena, 2019. **13**: p. 128-133.

[16] Aceves, S.M., et al., *Comparison of Cu, Ti and Ta interlayer explosively fabricated aluminum to stainless steel transition joints for cryogenic pressurized hydrogen storage.* International Journal of Hydrogen Energy, 2015. **40**(3): p. 1490-1503.

[17] Mastanaiah, P., et al., *An investigation on microstructures and mechanical properties of explosive cladded C103 niobium alloy over C263 nimonic alloy.* Journal of Materials Processing Technology, 2014. **214**(11): p. 2316-2324.

[18] Corigliano, P., et al., *Full-field analysis of AL/FE explosive welded joints for shipbuilding applications.* Marine Structures, 2018. **57**: p. 207-218.

[19] Crossland, B., F. McKee, and A. Szecket, *An experimental investigation of explosive welding parameters*, in *High-Pressure Science and Technology*. 1979, Springer. p. 1837-1845.

[20] Cowan, G., O. Bergmann, and A. Holtzman, *Mechanism of bond zone wave formation in explosion-clad metals*. Metallurgical and Materials Transactions B, 1971. **2**(11): p. 3145-3155.

[21] Wang, Y., et al., *Numerical simulation of explosive welding using the material point method*. International Journal of Impact Engineering, 2011. **38**(1): p. 51-60.

[22] Bondar, M. and V. Ogolikhin, *Plastic deformation in the joint zone with cladding by explosion*. Combustion explosion and shock waves, 1985. **21**(2): p. 266-270.

[23] Kriventsov, A. and V. Sedykh, *The role of plastic deformation of metal in the weld zone in explosion welding*. Fiz Khim Obrab Mater, 1969. **1**: p. 132-141.

[24] Krupin, A., et al., *Explosive Deformation of Metals*. Metallurgiya, Moscow, 1975.

[25] Gul'Bin, V. and A. Kobelev, *Plastic deformation of metals in explosion welding*. Welding international, 1999. **13**(4): p. 306-309.

[26] Borchers, C., et al., *Microstructure and mechanical properties of medium-carbon steel bonded on low-carbon steel by explosive welding*. Materials & Design, 2016. **89**: p. 369-376.

[27] Sabirov, I., M.Y. Murashkin, and R. Valiev, *Nanostructured aluminium alloys produced by severe plastic deformation: New horizons in development*. Materials science and engineering: A, 2013. **560**: p. 1-24.

[28] Gloc, M., et al., *Microstructural and microanalysis investigations of bond titanium grade1/low alloy steel st52-3N obtained by explosive welding*. Journal of Alloys and Compounds, 2016. **671**: p. 446-451.

[29] Bazarnik, P., et al., *Mechanical and microstructural characteristics of Ti6Al4V/AA2519 and Ti6Al4V/AA1050/AA2519 laminates manufactured by explosive welding*. Materials & Design, 2016. **111**: p. 146-157.

[30] Sherpa, B.B., et al., *Neuro-Fuzzy Technique for Micro-hardness Evaluation of Explosive Welded Joints*. Transactions of the Indian Institute of Metals, 2020. **73**(5): p. 1287-1299.

[31] Sherpa, B.B., et al., *Study of the Explosive Welding Process and Applications*, in *Advances in Applied Physical and Chemical Sciences-A Sustainable Approach*. 2014, Krishi Sanskriti: New Delhi. p. 33-39.

[32] Blazynsky, T., *Explosive forming, welding and compaction*. 1983, Elsevier Science, New York.

[33] Crossland, B. and J. Williams, *Explosive welding*. Metallurgical Reviews, 1970. **15**(1): p. 79-100.

[34] Abrahamson, G.R., *Permanent periodic surface deformations due to a traveling jet*. Journal of Applied Mechanics, 1961. **83**: p. 519-528.

[35] Loureiro, A., et al., *Effect of explosive mixture on quality of explosive welds of copper to aluminium*. Materials & Design, 2016. **95**: p. 256-267.

[36] Mendes, R., J. Ribeiro, and A. Loureiro, *Effect of explosive characteristics on the explosive welding of stainless steel to carbon steel in cylindrical configuration*. Materials & Design, 2013. **51**: p. 182-192.

[37] Manikandan, P., et al., *Control of energetic conditions by employing interlayer of different thickness for*

explosive welding of titanium/304 stainless steel. Journal of materials processing technology, 2008. **195**(1-3): p. 232-240.

[38] Bahadur Sherpa, B., et al., *Low Velocity of Detonation Explosive Welding (LVEW) Process for Metal Joining.* Propellants, Explosives, Pyrotechnics. 2020;**45**(10):1554-1565

[39] Hanliang, L., et al., *Joining of Zr60Ti17Cu12Ni11 bulk metallic glass and aluminum 1060 by underwater explosive welding method.* Journal of Manufacturing Processes, 2019. **45**: p. 115-122.

[40] Arab, A., et al., *Joining AlCoCrFeNi high entropy alloys and Al-6061 by explosive welding method.* Vacuum, 2020. **174**: p. 109221.

[41] Tanaka, S., A. Mori, and K. Hokamoto, *Welding of Sn and Cu plates using controlled underwater shock wave.* Journal of Materials Processing Technology, 2017. **245**: p. 300-308.

[42] Mori, A., M. Nishi, and K. Hokamoto, *Underwater shock wave weldability window for Sn-Cu plates.* Journal of Materials Processing Technology, 2019. **267**: p. 152-158.

[43] Tao, C., et al., *Microstructure and mechanical properties of Cu/CuCrZr composite plates fabricated by explosive welding.* Composite Interfaces, 2020: p. 1-12.

[44] Kaya, Y. and G. Eser, *Production of ship steel—titanium bimetallic composites through explosive cladding.* Welding in the World, 2019. **63**(6): p. 1547-1560.

[45] Durgutlu, A., B. Gülenç, and F. Findik, *Examination of copper/stainless steel joints formed by explosive welding.* Materials & design, 2005. **26**(6): p. 497-507.

[46] Kahraman, N., B. Gülenç, and F. Findik, *Joining of titanium/stainless steel by explosive welding and effect on interface.* Journal of Materials Processing Technology, 2005. **169**(2): p. 127-133.

[47] Manikandan, P., et al., *Underwater explosive welding of thin tungsten foils and copper.* Journal of Nuclear Materials, 2011. **418**(1-3): p. 281-285.

[48] Durgutlu, A., H. Okuyucu, and B. Gulenc, *Investigation of effect of the stand-off distance on interface characteristics of explosively welded copper and stainless steel.* Materials & Design, 2008. **29**(7): p. 1480-1484.

[49] Jandaghi, M.R., et al., *Microstructural Evolutions and its Impact on the Corrosion Behaviour of Explosively Welded Al/Cu Bimetal.* Metals, 2020. **10**(5): p. 634.

[50] Wittman, R. *Use of explosive energy in manufacturing metallic materials of new properties*. in *Proceedings of the Second International Symposium, Marianski Lazne, Czechoslovakia*. 1973.

[51] Cowan, G.R. and A.H. Holtzman, *Flow configurations in colliding plates: explosive bonding.* Journal of applied physics, 1963. **34**(4): p. 928-939.

[52] Vaidyanathan, P. and A. Ramanathan, *Computer-aided design of explosive welding systems.* Journal of materials processing technology, 1993. **38**(3): p. 501-516.

[53] Abrahamson, G.R., *Permanent periodic surface deformations due to a traveling jet.* journal of applied mechanics, 1961. **28**(4): p. 519-528.

[54] Zakharenko, I. and B. Zlobin, *Effect of the hardness of welded materials on the position of the lower limit of explosive welding.* Combustion, Explosion and Shock Waves, 1983. **19**(5): p. 689-692.

[55] Grignon, F., et al., *Explosive welding of aluminum to aluminum: analysis, computations and experiments.*

International Journal of Impact Engineering, 2004. **30**(10): p. 1333-1351.

[56] Acarer, M., B. Gülenç, and F. Findik, *Investigation of explosive welding parameters and their effects on microhardness and shear strength.* Materials & design, 2003. **24**(8): p. 659-664.

[57] Kacar, R. and M. Acarer, *Microstructure–property relationship in explosively welded duplex stainless steel–steel.* Materials Science and Engineering: A, 2003. **363**(1-2): p. 290-296.

[58] Kacar, R. and M. Acarer, *An investigation on the explosive cladding of 316L stainless steel-din-P355GH steel.* Journal of Materials Processing Technology, 2004. **152**(1): p. 91-96.

[59] Rybin, V., E. Ushanova, and N.Y. Zolotorevskii, *Features of misoriented structures in a copper-copper bilayer plate obtained by explosive welding.* Technical Physics, 2013. **58**(9): p. 1304-1312.

[60] Habib, M.A., et al., *Cladding of titanium and magnesium alloy plates using energy-controlled underwater three layer explosive welding.* Journal of Materials Processing Technology, 2015. **217**: p. 310-316.

[61] Fronczek, D., et al., *Structural properties of Ti/Al clads manufactured by explosive welding and annealing.* Materials & Design, 2016. **91**: p. 80-89.

[62] Lazurenko, D., et al., *Explosively welded multilayer Ti-Al composites: Structure and transformation during heat treatment.* Materials & Design, 2016. **102**: p. 122-130.

[63] Saravanan, S., K. Raghukandan, and K. Hokamoto, *Improved microstructure and mechanical properties of dissimilar explosive cladding by means of interlayer technique.* Archives of Civil and Mechanical Engineering, 2016. **16**: p. 563-568.

[64] Zhang, T., et al., *Microstructure evolution and mechanical properties of an AA6061/AZ31B alloy plate fabricated by explosive welding.* Journal of Alloys and Compounds, 2018. **735**: p. 1759-1768.

[65] Zhang, T.-T., et al., *Molecular dynamics simulations and experimental investigations of atomic diffusion behavior at bonding interface in an explosively welded Al/Mg alloy composite plate.* Acta Metallurgica Sinica (English Letters), 2017. **30**(10): p. 983-991.

[66] Carvalho, G., et al., *Microstructure and mechanical behaviour of aluminium-carbon steel and aluminium-stainless steel clads produced with an aluminium interlayer.* Materials Characterization, 2019. **155**: p. 109819.

[67] Kaya, Y., *Investigation of copper-aluminium composite materials produced by explosive welding.* Metals, 2018. **8**(10): p. 780.

[68] Paul, H., L. Lityńska-Dobrzyńska, and M. Prażmowski, *Microstructure and phase constitution near the interface of explosively welded aluminum/copper plates.* Metallurgical and Materials Transactions A, 2013. **44**(8): p. 3836-3851.

[69] Amani, H. and M. Soltanieh, *Intermetallic phase formation in explosively welded Al/Cu bimetals.* Metallurgical and Materials Transactions B, 2016. **47**(4): p. 2524-2534.

[70] Kaya, Y., *Microstructural, mechanical and corrosion investigations of ship steel-aluminum bimetal composites produced by explosive welding.* Metals, 2018. **8**(7): p. 544.

[71] Carvalho, G., et al., *Explosive welding of aluminium to stainless*

steel. Journal of Materials Processing Technology, 2018. **262**: p. 340-349.

[72] Sherpa, B.B., et al., *Examination of Joint Integrity in parallel plate configuration of explosive welded SS-Al combination.* Materials Today: Proceedings, 2017. **4**(2): p. 1260-1267.

[73] Sherpa, B.B., et al., *Interface Study of Explosive Welded AL-Steel Joint Using Ultrasonic Phased Array Technique*, in *31st International Symposium on Ballistics*. 2019, DEStech Publications, Inc. p. 2280-2290.

[74] Stivers, S. and R. Wittman, *Computer selection of the optimum explosive loading and weld geometry.* High Energy Rate Fabrication. University of Denver Research Institute, Colorado. 1975, 4. 2-4. 2. 16, 1975.

[75] Wylie, H. and B. Crossland. *Explosive cladding with thick flyer plates*. in *Proc. Conf. on The Use of High-Energy Rate Methods for Forming, Welding, and Compaction, Leeds*. 1973.

[76] Carpenter, S., R. Wittman, and R. Carlson, *Relationships of explosive welding parameters to material properties and geometries factors, proc. first int. conf. of the center for high energy forming, university of Denver*. 1967, June.

[77] Shribman, V. and B. Crossland. *An experimental investigation of the velocity of the flyer plate in explosive welding.* in *Second international conference of the center for high energy forming, Proceedings*. 1969.

[78] El-Sobky, H. and T. Blaznski, *Proc. 15th Int. MTDR Conf.* 1974.

[79] Findik, F., *Recent developments in explosive welding.* Materials & Design, 2011. **32**(3): p. 1081-1093.

Chapter 3

Influence of Annealed Aluminum Properties on Adhesion Bonding of Cold Sprayed Titanium Dioxide Coating

Noor Irinah Omar, Motohiro Yamada, Toshiaki Yasui and Masahiro Fukumoto

Abstract

It is well known that cold spraying ceramic materials can be difficult because cold spraying requires plastic deformation of the feedstock particles for adhesion to the substrate. The challenge lies in the difficulty of plastically deforming hard and brittle ceramic materials, such as TiO_2. Previous studies have reported the possibility of cold spraying thick pure TiO_2 but the bonding mechanism of cold sprayed TiO_2 is not fully understood. The factor like substrate condition as oxide film thickness and mechanical properties may also affect cold spray deposition but not fully understood in cold spraying ceramic. The aim of the present research is to investigate the correlation between the oxide thickness and substrate deformation with the adhesion strength of cold-sprayed TiO_2 coatings toward the bonding mechanism involved. Relevant experiments were executed using Al 1050, subjected to various annealing temperatures and cold-sprayed with TiO_2 powder. The results indicate a decreasing trend of coating adhesion strength with increasing annealed substrate temperature from room temperature to 400°C annealed. Metallurgical bonding is pronounced as bonding mechanism involved between TiO_2 particle and annealed 1050 substrate.

Keywords: cold spray, bonding mechanism, pure aluminum, titanium dioxide, annealed substrate, metallurgical bonding

1. Introduction

Cold spraying, also referred to as kinetic spraying, represents a relatively novel technique for material deposition that has been in development for over twenty years. The process involves the high-velocity acceleration of powder particles, typically in the 300–1200 m/s range, in a jet flow of supersonic velocity with the projection directed onto a substrate or a coating that has been pre-deposited at an absolutely solid state. Due to the impact at high velocity, intensive deformation of plastic manifests in the cold-sprayed particles or substrate, allowing the formation of a low oxidized cold-sprayed coating [1]. Since nanostructured TiO_2-anatase coatings have an extensive active surface and chemical stability, and are at a comparatively reasonable cost, they are utilized as functional material [2]. The crystalline framework of TiO_2 has a major impact upon its photocatalytic performance where TiO_2 in anatase stage supplies greater photocatalytic activity than in its rutile stage. At temperatures exceeding 900°C, which is above the melting point of TiO_2 (1908°C), the anatase stage irreversibly changes into the rutile stage. At temperatures above the melting point of TiO_2, the deposition of molten or semi-molten droplets form thermal impossible to avert the phase transformation of TiO_2 in thermal spray procedures [2]. Therefore, cold spraying is the best solution because it sprays below the material melting temperature.

In terms of the mechanism for bonding, numerous works have been conducted in the past 10 years to enhance our understanding of this phenomenon [3–14]. At present, the manifestation at the interface of adiabatic shear instability, (ASI) represents the most accepted bonding-mechanism theory, resulting from the high rate of strain as well as the intense local deforming that occurs through the process of depositing the particles. In the vicinity of adiabatic shear instability occurrence, adiabatic heating-induced thermal softening dominates the hardening through work, and therefore the metals' behavior is similar to viscous material and leads to extrusion from the interface, with the formation of an outward metal jet at the rim [4–5]. The presence of the metal jet with viscosity assists in the cleaning of the native-oxide film that is cracked and originates from the surfaces of particle or substrate, facilitating contact between the metals and therefore the occurrence of metallic bonding [6]. Two mechanisms widely perceived to dominate in terms of metallic bonding in cold spray are interlocking through mechanical means and bonding metallurgically.

It is interesting that cold spraying can also be used to deposited ceramic materials, although initially it appears impossible as cold spraying require plastic deformation to work. Previous studies have reported the possibility of cold spraying thick pure titanium dioxide, TiO_2 within range 300 μm [15] but the bonding mechanism of cold sprayed TiO_2 is not fully understood. Several experiment result regarding the bonding mechanism of cold spraying TiO_2 particles onto metal and ceramic substrates have been published. Toibah et.al reported as-synthesized TiO_2 powders that calcined at 200°C and 300°C showed the successful deposition of TiO_2 coating on the ceramic tile substrate by the cold spray method. The coating deposition occurred due to tamping effect through slipping of nanoparticles due to high impact during the particle collision. Mechanism responsible for the TiO_2 deposition is chemical bonding between TiO_2 particles and substrate or among particles during cold sprayed process may lead to coating formation [16]. Kliemann et al. used 3-50 μm TiO_2 agglomerates formed from 5 to 15 nm primary particles for the continuous coating on pure titanium, stainless steel, copper and aluminum alloy substrate. They identified ductile substrate that allow shear instability to happen as primary bonding mechanism between the particles and the substrate [17]. Gardon et al. claimed that the mechanism responsible for the TiO_2 deposition by the cold spraying process on the stainless-steel substrate is the chemical bonding between the particles and the substrate. They showed that the previous layer of titanium sub-oxide prepares the substrate with the appropriate surface roughness needed for the TiO_2 particle deposition. Moreover, the substrate composition is also important for the deposition because it can provide chemical affinities during the particle interaction after impact. The substrate hardness may also ease the interaction between the particles and the substrate [2].

Based on the above mention meaningful findings, it shown a lot of factor influenced toward bonding mechanism of cold sprayed TiO_2 coating and one of it is substrate surface. Previous studies reported the possibility of cold spraying thick pure TiO_2 but the bonding mechanism of cold sprayed TiO_2 is not fully understood. The factor like substrate condition as oxide film thickness and mechanical properties may also affect cold spray deposition but not fully understood in cold spraying ceramic. Therefore, this study investigated the effect of annealed substrate properties Al 1050 toward bonding mechanism of cold sprayed TiO_2.

2. Materials and methods

2.1 Process

In all coating experiments, cold-spraying equipment with a De-Laval 24TC nozzle (CGT KINETIKS 4000; Cold Gas Technology, Ampfing, Germany) was used. Nitrogen was used as the process gas with a 500°C operating temperature, and a 3 MPa pressure. The spray distance was 20 mm, with a process traverse speed of 10 mm/s. The coatings were deposited on grit-blasted annealed pure aluminum (Al 1050). The substrates were annealed with an electric furnace to preheat the grit-blasted substrate to four different temperatures respectively (i.e., 100°C, 200°C, 300°C and 400°C) before spraying. In all cases, the temperature of the annealed substrate during spraying was room temperature.

2.2 Materials

As a feedstock, we applied agglomerated TiO_2 powder (TAYCA Corporation, WP0097) containing a pure anatase crystalline structure with an average particle size of about 7.55 μm. The material chemical composition of substrates used is presented in **Table 1**.

2.3 Characterization

2.3.1 Tensile-strength testing

In accordance with JIS H 8402, specimens measuring Ø25 mm × 10 mm were used to assess the coatings' adhesion strength, given as the fracture load value measured by a universal testing machine (Autograph AGS-J Series 10 kN, Shimadzu). We measured the adhesion strength over an average of five specimens for each of the spraying conditions.

2.3.2 Coatings evaluation

A scanning electron microscope (SEM: JSM-6390, JEOL, Tokyo, Japan) was used to observe the TiO_2 coating's cross-sectional microstructures on annealed substrates. The observation sample of the TiO_2 coating was prepared by embedding a 25mmx10mm sample into a hardenable resin. The hardened sample embedded in hardened resin was ground with silica papers to a #3000 grit size and finally polished with 1 μm and 0.3 μm alumina suspension.

	Al	Fe	Si	Cu	Mn	Mg	Zn	Ti	Other total
Al 1050	Bal	0.40	0.25	0.05	0.05	0.05	0.05	0.05	0.03

Table 1.
Material chemical composition (wt.%).

Measured regions	Al 2p,	O1s,
Measured X-Ray output [W]	10	
Probe diameter[μm]	50	
Time per step [ms]	30	
Pass energy	140	
Cycle	30	

Table 2.
XPS parameters for substrate oxide layer analysis.

2.3.3 Micro-Vickers hardness

To investigate the relationship between the annealed substrate surface hardness and the adhesion strength of the TiO_2 coating on the annealed substrate, the substrate hardness was measured using an HMV-G micro-Vickers hardness tester (Shimadzu). The measurement showed a hardness of HV 0.1; the test load on the cross section was 98.07[mN]. The final micro-hardness value was the average of 5 tests taken at approximately the same points for each substrate.

2.3.4 Substrate oxide evaluation

X-ray photoelectron spectroscopy (XPS) is a versatile surface analysis technique used for compositional and chemical state analyses. In this study, XPS analysis (ULVAC-PHI, PHI Quantera SXM-CI) using a monochromatic Al Kα source (15 mA, 10 kV) was performed. Wide (0–1000 eV) and narrow scans of Al 2P and O1s for different annealed substrates were collected. The measured binding energies were then corrected with C 1 s at 285.0 eV. When pre-sputtering to clean the surface was performed, the sample surface was reduced and the measurements were affected, so XPS analysis was performed without pre-sputtering. **Table 2** shows the XPS analysis conditions for substrates oxide analysis.

2.3.5 Wipe test

A CGT Kinetiks 4000 cold-spray system (Cold Gas Technology, Ampfing, Germany) with a custom-made suction nozzle was used to perform the wipe test and coating using TiO_2 powder onto the annealed 400°C pure aluminum substrates. The wipe test was conducted to study the deformation behavior of a single particle on this substrate. Prior to deposition, substrate was ground and polished until a mirror finish surface was obtained. The process gas temperature and pressure used were 500°C and 3 MPa, respectively. Nitrogen was used as the process gas. The distance between the exit of the nozzle and the substrate was fixed at 20 mm. The traverse speed of the process was 2000 mm/s. Prior to spraying, the substrates were rinsed with acetone. An FEI Helios Dual Beam 650 field emission SEM (FESEM) and focused ion beam (FIB) microscope was used to investigate the single particle TiO_2 deposition onto mirror polish annealed substrate.

3. Results and discussion

3.1 Strength of adhesion

The adhesion strength of the cold-sprayed TiO_2 coating on annealed Al 1050 pure aluminum are shown in **Figure 1**. The TiO_2 coating on the annealed soft substrates showed a decreased trend of adhesion strength from room temperature to 400°C, with values from 3.88 to 2.05 MPa. The increase of annealed substrate temperature also showed a thicker surface oxide on substrate surface.

3.2 SEM cross-section microstructure of TiO_2 coatings on annealed Al 1050 substrates

Figure 2a–e shows the TiO_2 coating cross-sectional area on Al 1050 for various substrate annealing temperatures. Under all condition figures, the figures show a dense coating with a thickness in the range of 200 to 300 μm, indicating that a critical velocity was reached for this soft material. This suggests that the TiO_2 coating adhered well to the annealed Al 1050 substrate from room temperature to 400°C annealing.

We can categorize the cold-spraying procedure into two stages: (1) adhesion and (2) cohesion bonding. Adhesion or the formation of the interface between the substrate and particle is the first stage. The annealed substrates can clearly implement this stage, which forms the first coating layer, particularly for the soft material, Al 1050.

3.3 Substrate Vickers microhardness

Figure 3 shows the annealing substrate hardness of Al 1050 from room temperature to 400°C. The pure aluminum, Al 1050 showed a decreasing trend from 45.36Hv for room temperature to 27.70Hv for 400°C annealing.

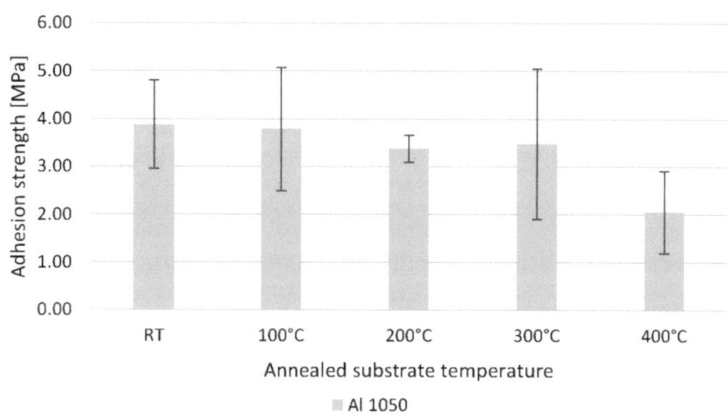

Figure 1.
Adhesion strength of the TiO_2 coating on annealed Al 1050.

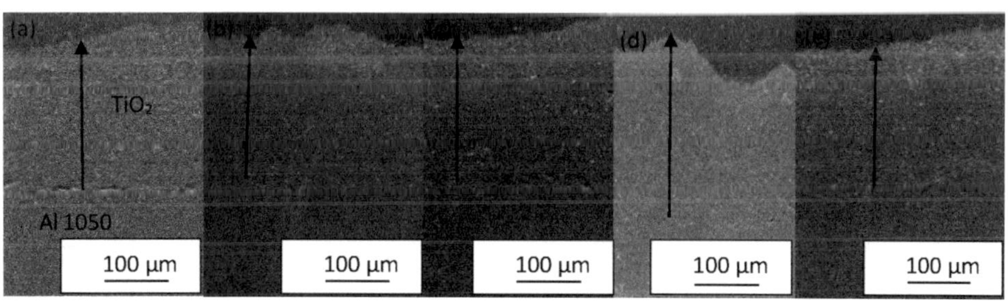

Figure 2.
Cross-section microstructure of TiO_2 coatings on Al 1050. (a) Room temperature; (b) annealed 100°C; (c) annealed 200°C; (d) annealed 300°C and (e) annealed 400°C.

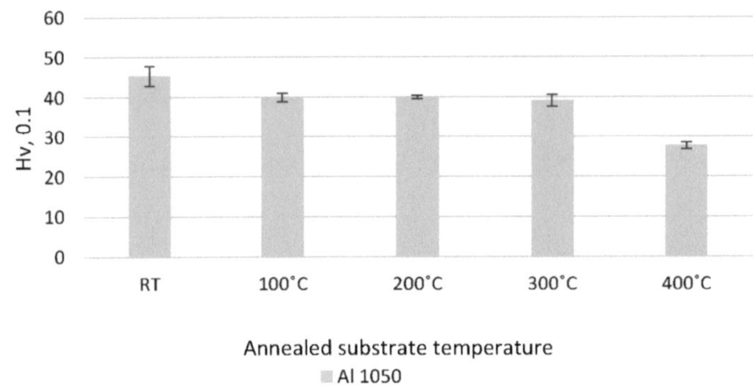

Figure 3.
Annealed substrate microhardness of Al 1050 from room temperature to 400°C- annealed.

This trend appears because pure materials experienced recrystallization process at 400°C and slow cooling in the electric furnace; allow the grains to growth larger and reduce the substrate hardness. Based on decrease trend shown by adhesion strength TiO_2 coating on annealed pure aluminum, substrate deformation (mechanical anchoring) due to TiO_2 particle impacting during cold spraying is not the main factors that influence adhesion bonding between TiO_2 and annealed pure aluminum.

3.4 Depth profile of the oxide layer on Al 1050 substrate

The results of the depth analysis of room temperature substrate and annealed 400°C by X-ray photoelectron spectroscopy for Al 1050 substrates is shown accordingly in **Figure 4**. The composition as a function of depth can be analyzed by in-situ argon ion beam sputtering, found on most surface analytical equipment.

Al 1050 as shown by **Figure 4(a)** and **(b)**, shown that the atomic composition of Oxygen in the deepest part of the oxide layer increases as the annealing substrate temperature increases from RT to 400°C. This indicate that the oxide layer of pure aluminum grows thicker as the annealing substrate temperature is increased. According to W. Ya Li et.al stated that the increasing of oxide film thickness, it will need more kinetic energy to break up and extrude the oxide film, thus a higher particle velocity is needed for bonding. In other words, the effective bonding area is decreased under the same particle impact conditions [18]. It could explain the decreased trend of adhesion strength TiO_2 coating on 400°C annealed pure aluminum because the particle velocity was constant in all condition in this experiment.

3.5 Oxide layer composition

X-Ray photoelectron spectroscopy (XPS) is a versatile surface analysis technique that can be used for compositional and chemical state analysis. In this experiment, total of 30 cycles is involved and to plot binding energy graph versus intensity to study the chemical state changes, outermost surface referring to the 1st cycle data, mid-layer referring to the 15th cycle data and deepest part referring to the 30th cycle data for each substrate conditions.

The peak position of the aluminum oxide layer, Al_2O_3 on outermost surface at 74.8 eV and aluminum metal at approximately 72.8 eV was prominently present in the mid-layer and deepest part for the room-temperature substrate Al 1050 shown by **Figure 5(a)**. The peak position of the O1s shown by **Figure 5(b)** also support the present of aluminum oxide layer, Al_2O_3 on outermost surface at 531.3 eV and

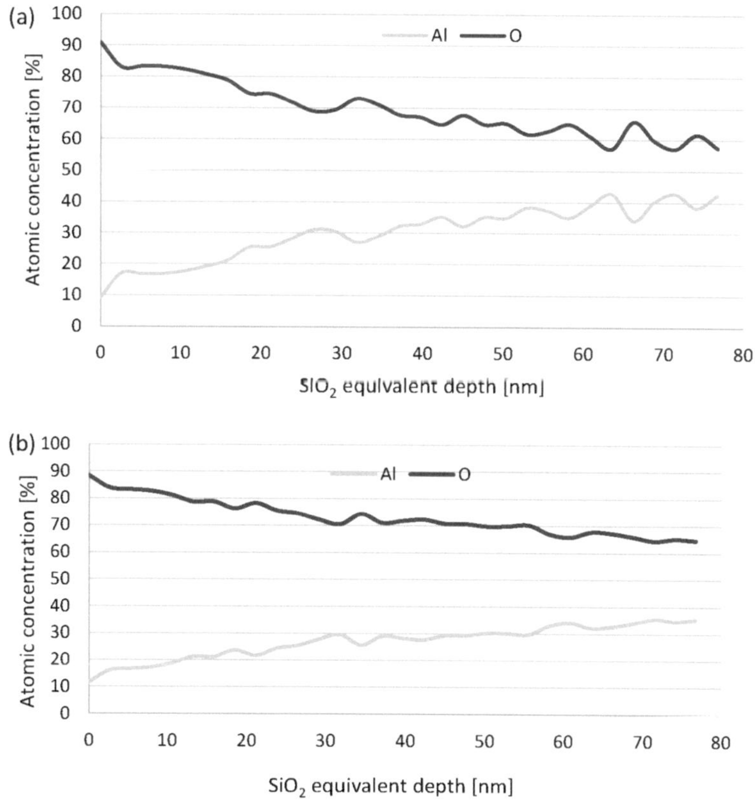

Figure 4.
Depth profile analysis of Al 1050 (a) room temperature (b) 400°C annealed.

slightly presented of aluminum in hydroxide on mid-layer and deepest part given by 532.4 eV, supported that substrate mostly Al metal state for mid layer and deepest part at room temperature substrate. Meanwhile, at the annealing temperature of 400°C for Al 1050 substrate, **Figure 6(a)**, the peak position of aluminum oxide layer, which was at approximately 74.8 eV, was detected at the outermost surface, mid-layer, and deepest part of the substrates. O1s supported that by the present of aluminum in hydroxide/oxyhydroxide state at 532.4 eV [19]. This indicate Al 1050 substrate experienced chemical composition changes from Al metal state in room temperature substrate to aluminum in hydroxide condition in 400°C annealed.

3.6 FIB splat TiO_2 particle on 400°C annealed Al 1050

Figure 7 shows the top view of TiO_2 particles on 400°C-annealed Al 1050. This wipe test was conducted to further understand the bonding mechanism of TiO_2 particle on annealed Al 1050 substrate. Only 400°C-annealed Al 1050 was selected because it had the highest oxide thickness. The results obtained revealed that the TiO_2 particle were found unchanged after the collision and the substrate surface of the 400°C-annealed Al 1050 experienced a deformation due to impacting during the cold-spraying process, as shown by **Figure 7**. Soft substrates such as aluminum, the TiO_2 particles were impacting the surface with minimal deformation, and the particles rebounded after the impact, leaving craters on the surface of the substrate. This shown by arrow in the **Figure 7**. K.-R. Ernst et al. [20] also mentioned that for soft substrate, impact energy that was generated during the spraying process was also used for deformation of the substrate. Since 400°C

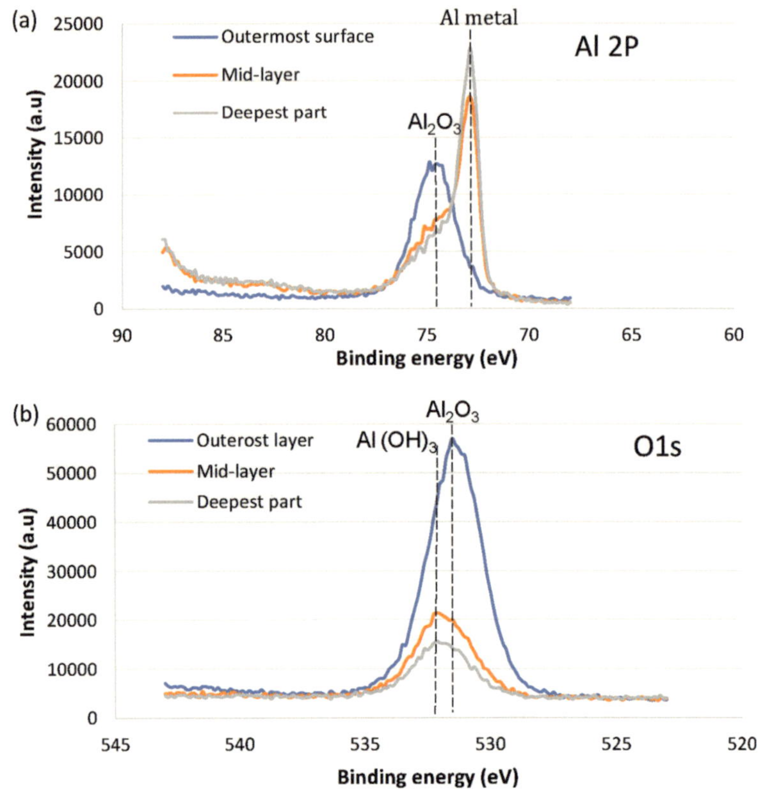

Figure 5.
XPS spectra of Al 2P for (a) room temperature and O1s for (b) room temperature.

annealed aluminum has lower hardness due to recrystallization, the depth of the craters was deeper and experienced heavy damage, as seen in the SEM image in **Figure** 7.

Figure 8 shown a cross-section of a single particle of TiO_2 on 400°C annealed 1050. Referring to **Figure 8**, since the shear instability starts at a position away from the bottom center of a TiO_2 particle, the bottom region of such a deposited particle can be divided into three regions along the particle - substrate boundary: (i) the particle jetted out region,(J) generated by the severe shear plastic strain induced by adiabatic shear instability (ASI); (ii) the well-bonded region (B); where the particle and the substrate are intimately bonded and (iii) the rebound region (R) where the shear instability did not occur and the accumulated elastic energy from the impact of a sprayed particle detached the particle from the substrate. At the boundary of B and R, ASI is accompanied by severe shear stress, and an abnormal increases in temperature can easily expel the particles, and consequently the oxide covering the surface of particle or substrate can be broken and removed [3, 4, 11, 21–24].

The adhesion strength of the TiO_2 coating on annealed Al 1050, showed a decreased trend as the annealed substrate temperature is increased from room temperature to 400°C annealed. This indicate that substrate deformation or mechanical anchoring is not one of factors that influence the adhesion bonding for annealed Al 1050 with TiO_2 at an annealing temperature of 400°C. Referring to arrow indicate in **Figure 8**, there were a gap observe between deposited TiO_2 particle and 400°C annealed 1050 substrate at rebound, R and bonded, B region. This result indicate that even severe plastic deformation is occurred due to TiO_2 impacting during cold spraying, thin amorphous oxide will remain between particle-substrate [25].

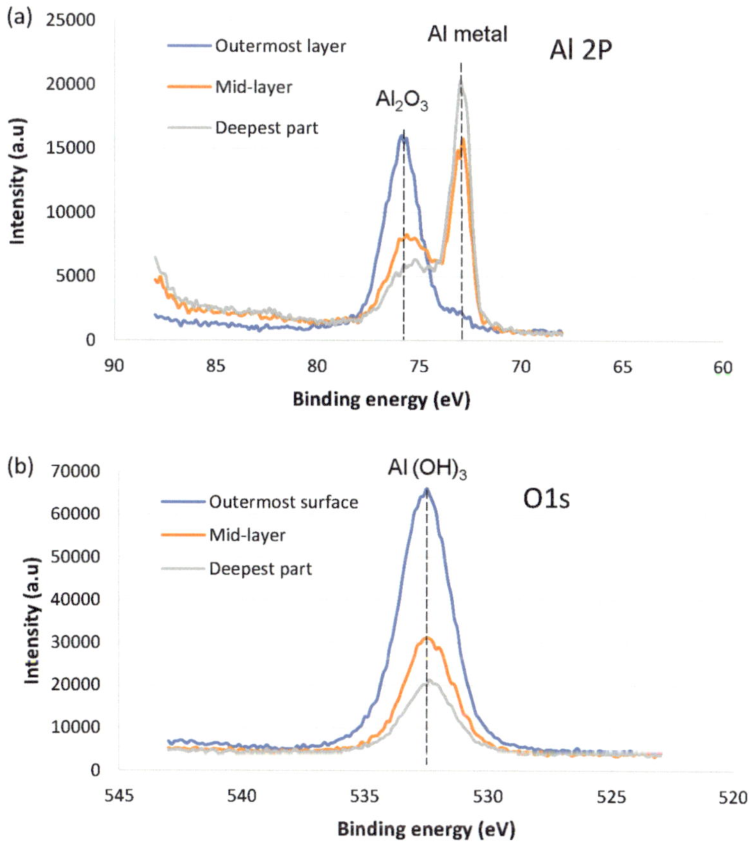

Figure 6.
XPS spectra of Al 2P for (a) 400°C annealed and O1s for (b) 400°C annealed.

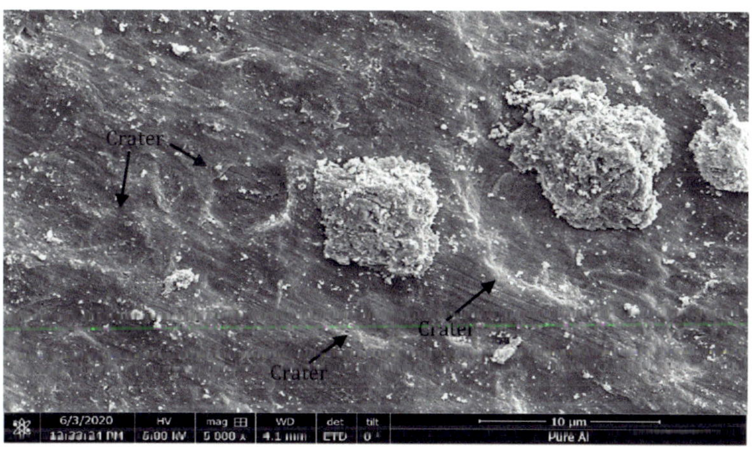

Figure 7.
Top view of TiO_2 particles on 400C annealed Al 1050.

Moreover, due to thicker oxide on 400°C annealed Al 1050 or the fresh surfaces of metallic such as aluminum may be able to be re-oxidized due to their high reactivity with oxygen in air during deformation [25], it prevent bonding to form between deposited TiO_2 particle and deform substrate surface of 400°C annealed Al 1050. This may explain that bonding mechanism involved between TiO_2 coating and

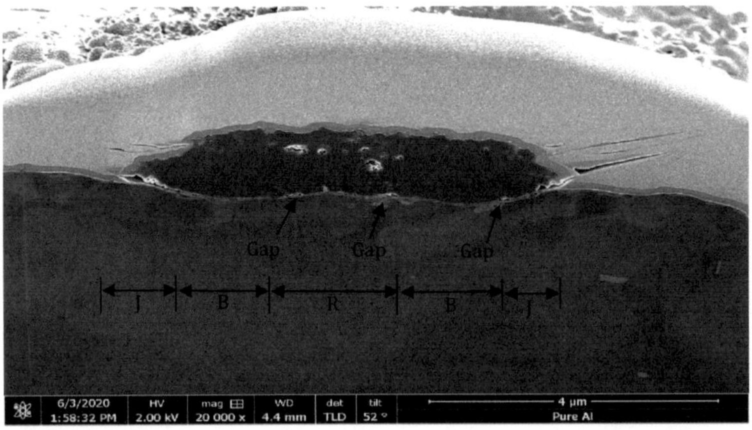

Figure 8.
FIB cross-section of a single particle of TiO_2 on 400°C-annealed Al 1050. J indicates the jetted-out region; B is the bonded region; R is the rebound region.

400°C annealed 1050 is newly formed substrate surface that free from oxide with TiO_2 particle which is similar with the cold spraying of metallic materials that known as metallurgical bonding.

4. Conclusions

This study investigated the correlation between the adhesion strength of cold-sprayed TiO_2 on Al 1050 pure aluminum—annealed at temperatures ranging from room temperature to 400°C.

1. The adhesion strength of TiO_2 coating showed a decreased trend from 3.88 MPa to 2.05 MPa as annealed substrate temperature increased from room temperature to 400°C.

2. The oxide film thickness had an influence on the TiO_2 deposition process because thick oxide film inhibits formation of a new surface.

3. The main bonding mechanism is metallurgical bonding which is newly form substrate surface of annealed Al 1050 that oxide free to TiO_2 particle.

Acknowledgements

We also acknowledge the Interface & Surface Fabrication lab, Majlis Amanah Rakyat, MARA and the Universiti Teknikal Malaysia Melaka, UTeM for Noor Irinah's PhD scholarship.

Funding

This research is supported by JSPS KAKENHI Grant Number JP17K06857 and was partially carried out at the Cooperative Research Facility Center at Toyohashi University of Technology.

Author details

Noor Irinah Omar[1,2]*, Motohiro Yamada[1], Toshiaki Yasui[1] and Masahiro Fukumoto[1]

1 Department of Mechanical Engineering, Toyohashi University of Technology, Tempaku-Cho, Toyohashi, Aichi, Japan

2 Fakulti Teknologi Kejuruteraan Mekanikal dan Pembuatan, Universiti Teknikal Malaysia Melaka, Durian Tunggal, Melaka, Malaysia

*Address all correspondence to: noor.irinah.binti.omar.dj@tut.ac.jp; nooririnah@utem.edu.my

References

[1] V.K.Champage, editors. The cold spray materials deposition process fundamentals and applications: Woodhead publishing; 2007.1-7p. DOI:10.1533/9781845693787.1

[2] M. Gardon, C. Ferna'ndez-Rodrı'guez, D. Garzo'n Sousa, J.M. Don˜a-Rodrı'guez, S. Dosta, I.G. Cano, and J.M. Guilemany. Photocatalytic activity of nanostructured anatase coatings obtained by cold gas spray. J. Therm. Spray Technol.2014; 23:1135-1141. DOI: 10.1007/s11666-014-0087-0

[3] H.Assadi, F.Gärtner,T.Stoltenhoff and H.Kreye. Bonding mechanism in cold spraying.Acta Mater. 2003;51: 4379-4394. DOI: 10.1016/S1359-6454(03)00274-X

[4] M.Grujicic, C.L.Zhao, W.S.DeRosset and D.Helfritch. Adiabatic shear instability based mechanism for particle/substrate bonding in the cold-gas dynamic spray process.Mater and Design. 2004;25: 681-688. DOI: 10.1016/j.matdes.2004.03.008

[5] T.Schmidt, F.Gärtner, H.Assadi and H.Kreye : Development of a generalized parameter window for cold spray deposition. Acta Mater.2006;54:729-742. DOI: 10.1016/j.actamat.2005.10.005

[6] R.C.Dykhuizen, M.F.Smith, D.L.Gilmore, R.A.Neiser,X.Jiang and S.Sampath. Impact of high velocity cold spray particles. J.Therm. Spray.Technol.1999;8:559-564. DOI: 10.1361/105996399770350250

[7] M.Grujicic, J.R.Saylor, D.E.Beasley, W.S.DeRosset and D.Helfritch. Computational analysis of the interfacial bonding between feed-powder particles and the substrate in the cold-gas dynamic-spray process. Appl.Surf.Sci. 2003; 219: 211-227.DOI: 10.1016/S0169-4332(03)00643-3

[8] S.Barradas, R.Molins, M.Jeandin, M.Arrigoni, M.Boustie, C.bolis, L.Berthe and M.Ducos. Application of laser shock adhesion testing to study of the interlamellar strength and coating-substrate adhesion in cold-sprayed copper coating of aluminum. Surf and Coat Tech. 2005; 197:18-27. DOI: 10.1016/j.surfcoat.2004.08.222

[9] S.V.Klinkov, V.F.Kosarev and M.Rein. Cold spray deposition: significance of particle impact phenomena. Aero Sci and Tech.2005;9: 582-591.DOI: 10.1016/j.ast.2005.03.005

[10] T.S.Prince,P.H.Shipway, D.G.McCartney, E.Calla and D.Zhang. A method for characterizing the degree of inter-particle bond formation in cold sprayed coatings. J.Therm Spray Techol.2007;16: 566-570.DOI:10.1007/s11666-007-9070-3

[11] G.Bae, Y.Xiong, S.Kumar, K.Kang and C.Lee. General aspects of interface bonding in kinetic sprayed coatings. Acta Mater.2008;56: 4858-4868.DOI: 10.1016/j.actamat.2008.06.003

[12] G.Bae, S.Kumar, S.Yoon, K.Kang, H.Na, H.Kim and C.Lee. Bonding features and associated mechanisms in kinetic sprayed titanium coatings. Acta Mater. 2009; 57:5654-5666. DOI: 10.1016/j.actamat.2009.07.061.

[13] T.Hussain, D.McCartney, P.Shipway and D.Zhang. Bonding mechanism in cold spraying: the contributions of metallurgical and mechanical components. J.Thermal Spray Technology. 2009;18: 364-379.DOI: 10.1007/s11666-009-9298-1

[14] S.Guetta, M.H.berger, F.Borit, V.Guipont, M.Jeandin, M.Boustie, Y.Ichikawa, K.Sakaguchi and K.Ogawa. Influence of particle velocity on adhesion of cold-sprayed splats.

J.Therm.Spray Technol.2009;18:331-342. DOI: 10.1007/s11666-009-9327-0

[15] M.Yamada, H.Isago,H.Nakano and M.Fukumoto. Cold spraying of TiO2 photocatalyst coating with nitrogen process gas. J. Therm. Spray Technol. 2010; 19:1218-1223. DOI: 10.1007/s11666-010-9520-1

[16] A.R.Toibah, K.Takahashi, M.Yamada and M.Fukumoto. Effect of powder calcination on the cold spray titanium dioxide coating.Mater. Trans.2016;57:1345-1350. DOI: 10.2320/matertrans.T-M2016817

[17] J.O.Kliemann, H.Gutzmann, F.Gartner, H.Hubner, C.Borchers and T.Klassen. Formation of cold-sprayed ceramic titanium dioxide layers on metal surfaces. J.Therm Spray Technol.2010;20:292-298.DOI: 10.1007/s11666-010-9563-3

[18] W-Ya Li, Chang-Jiu Li and Hanlin Liao. Significant influence of particle surface oxidation on deposition efficiency, interface microstructure and adhesive strength of cold-sprayed copper coatings. Appl. Surf. Sci. 2010; 256: 4953-4958. DOI: 10.1016/j.apsusc.2010.03.008

[19] N.Kumar and K.Biswas. Cryomilling: An environment friendly approach of preparation large quantity ultra refined pure aluminium nanoparticles. J.Mater.Research and technol. 2019;8:63-74. DOI: 10.1016/j.jmrt.2017.05.017

[20] K.-R. Ernst, J. Braeutigam, F. Gaertner and T. Klassen. Effect of substrate temperature on cold-gas-sprayed coatings on ceramic substrate.J.Therm.Spray Technol.2012;22: 422-432.DOI: 10.1007/s11666-012-9871-x

[21] Y.Xie, M-P.Planche, R.Roaelison,H. Liao,X.K.Suo and P.Herve. Effect of substrate preheating on adhesion strength of SS 316L cold spray coatings. J.therm Spray Technol.2016;25:123-130. DOI: 10.1007/s11666-015-0312-5

[22] S. Yin, X Wang, X Suo, H Liao, Z Guo, W Li and C Coddet. Deposition behavior of thermally softened copper particles in cold spraying. Acta Mater. 2013,61, 5105-5118.DOI: 10.1016/j,surfcoat.2014.11.009

[23] S.Yin, X.f.Wang, W.Y.Li and H.E.Jie. Effect of substrate hardness on the deformation behavior of subsequently incident particles in cold spraying. Appl. Surf.Sci. 2011;257:7560-7565. DOI: 10.1016/j.apsusc.2011.03.126

[24] K.Kim,W.Li and X.Guo. Detection of oxygen at the interface and its effect on strain, stress and temperature at the interface between cold sprayed aluminum and steel substrate. Appl. Surf.Sci. 2015;357:1720-1726. DOI: 10.1016/j.apsusc.2015.10.022

[25] K.H.Kim and S.Kuroda. Amorphous oxide film formed by dynamic oxidation during kinetic spraying of titanium at high temperature and its role in subsequent coating formation. Scr. Mater. 2010;63: 215-218.DOI: 10.1016/j.scriptamat.2010.03.061

Chapter 4

Anisotropic Mechanical Properties of 2-D Materials

Qiang Li

Abstract

While prior reviews and research articles focused on the various synthetic routes and microstructural controls of 2D nanomaterials as well as their functional applications, this chapter discloses the anisotropic behaviors of 2D materials and puts emphasis on the mechanical anisotropy of three distinct 2D materials, namely graphene, MoS_2 and Al alloy coating, representative of carbon, inorganic and metallic 2D crystalline materials. Except for the relatively low interlayer cohesive stress, the in-plane anisotropy of the former two materials classes is subjected primarily to the hexagonal structure of the unit cells of the graphene and MoS_2. The anisotropy of metallic thin films with high-density grain boundaries with preferential directionality, rendered by the non-equilibrium synthetic methods, results from both the conventional Taylor factor and the directionality of the grain boundaries. Despite 2D materials' wide spectrum of applications, such as electronics, energy devices, sensors, coating etc., the mechanical anisotropy could be critical for certain mechanical applications, such as friction, and provide instructions on the durability, reliability and property optimization in the various applications of different 2D materials.

Keywords: anisotropy, mechanical behaviors, 2D materials, metallic materials, non-metallic materials

1. Introduction

As time passes, the advancement of nanotechnology has spread to all fronts. The concept states that at least two dimensions that construct nanomaterials fall between 1 and 100 nm. Nanomaterials are classified as zero (0D), one- (1D), two- (2D), and three-dimensional (3D) nanostructures. The nanoscale has unprecedented attributes that fundamentally alter materials' properties. Since K. Novoselov et al. successfully mechanically exfoliated a single layer of graphene off the graphite in 2004 [1], extensive efforts and progress have been made on the synthesis and applications of graphene and various 2D nanomaterials in resemblance to graphene nanostructure, including transition metal dichalcogenides (TMDs), hexagonal boron nitride (BN), and perovskites, just to name a few. They have lateral extension but their individual layer is merely a single or few atoms thick. Hence, they have characteristics like electron confinement and anisotropy in various properties manifested in two dimensions, while they possess extended interlayer spacing for active kinetic and physicochemical events, which has attracted broad research interests on their physicochemical, electrochemical, electronic and mechanical properties. For metallic materials, metallic thin films/coatings have 2D extension but limited thickness. The 2D materials selected to represent each materials family are crystalline materials and, in general, possess crystal anisotropy in their mechanical behaviors and even functional property. The single crystal face-centered cubic (FCC) structure is taken as an example. It is well known that the FCC single crystals have crystal anisotropy determined by the Schmid factor that associates the loading direction to the load resolved on the specific slip system [2]. Because the glide of the dislocations is favored on the slip systems subjected to a larger Schmid factor, plastic anisotropy manifests in the form of different cellular substructure made of dislocation walls. As we alleged, the non-metallic 2D nanomaterials have extended interlayer spacing and metallic thin films, fabricated primarily by ultrahigh vacuum techniques and electrodeposition, feature high-density directional grain boundaries (GBs) and preferential texture. As a result, anisotropy in 2D materials is prone to deviate from that of the bulk crystals, and plays substantial roles in their mechanical applications and the reliability of the apparatuses and devices with 2D materials as components or building blocks, but has not been put emphasis on as much as their functional properties and synthesis. Prior to comprehension toward the mechanical anisotropy of 2D materials, their general microstructural features and applications are first set forth so as to better grasp the anisotropy in their mechanical response to external stimuli and its importance in their functional and engineering applications.

1.1 The general microstructural features and applications of 2D materials

1.1.1 Non-metallic 2D materials

Graphene is a typical 2D carbon allotrope and a monolayer of graphene, with a thickness of 0.335 nm, has a hexagonal honeycomb structure. This 2D nanomaterial is remarkably electric and thermal conductive and is equipped with the promising quantum Hall effect. Moreover, the pristine graphene possesses an elastic modulus of ~1 TPa and a mechanical quality factor of 10^4 at an elevated temperature of 5 K. Despite similar sheet-like nanostructures, 2D nanomaterials made of inorganic compounds can render intriguing properties and versatility due to their more complex compositions. In contrary to chemically inert graphene with no intrinsic bandgap, MoS_2 with layered structure is one transition metal dichalcogenide. MoS_2 has been often synthesized using chemical vapor deposition (CVD) and its reaction principle involves first the transformation of solid-state MoO_3 and sublimed surfur to gas state and then the mixed gases driven by argon caused the formation of gas-phase $MoO_{3\text{-}x}$ and MoS_2 and eventually the formation of solid-state MoS_2, the reactions of which is expressed as [3]:

$$MoO_3(g) + \left(\frac{x}{2}\right)S(g) \leftrightharpoons MoO_{3-x}(g) + \left(\frac{x}{2}\right)SO_2 \tag{1}$$

$$MoO_{3-x}(g) + (7-x)S(g) \leftrightharpoons MoS_2(g) + \left(\frac{3-x}{2}\right)SO_2. \tag{2}$$

MoS_2 structure is comprised of two layers of closely packed S atoms layers sandwiching a layer of Mo atoms and it features strong covalent bonding as a result of the Mo-S interaction and the Van der Waals force between S layers. This leads to the comparably facile kinetic transportation of ions and even molecules through

S-Mo-S layered structure [4]. Hence, it has drawn enormous attention for its potent applications in energy storage and conversion, such as photocatalysis for the pollutant degradation and biosensors, just to name a few. In addition, MoS_2 has a good tunability toward its band gap, which offers high flexibility in property customization and optimization. At the same time, MoS_2 manifests comparable physical attributes when compared to graphene, including high charge carrier mobility and superb wear properties. Compared to hexagonal structure, other structurally complex 2D materials, such as arsenic trisulfide (As_2S_3), also showed mechanical anisotropy [5]. A unit cell of As_2S_3 consists of two layers inverted with respect to a symmetry center and is defined by 20 atoms in contrast to two for graphene and three for MoS_2.

1.1.2 Metallic thin films

Now, we turn our attention to a different 2D materials family, i.e. metallic coating and thin films. The protective coatings, from an engineering point of view, are essential as to apparatus maintenance and the enhanced equipment safety and lifespan. One application of metallic or their composite coating is to prevent corrosion. Coatings should render compatibility with parental materials and operate at extreme atmospheres, such as high temperature and corrosive conditions. Metallic coatings either provide passive protection by forming a barrier of oxides or offer active protection obtained through the adsorption of chemical inhibitors [6]. Metallic coatings as biomaterials are potent components in body implants and they ought to possess superb mechanical behaviors and biocompatibility, and high corrosion resistance, while they are required to release minimal metallic ions to avoid the toxicity. Ti, NiTi, Pt and 316 L austenitic stainless steel are often implemented. Furthermore, various metals with unique characteristics are used in the applications of thin film optics, such as surface plasmon generation and optoelectronics. Metallic thin films are frequently applied onto ceramic matrices, rendering high-quality broadband reflective finishing highly desired to control over the directionality of the laser beam. In addition, Cu is commonly utilized as an interconnect material and serves as thin conductive layers ensuring the adhesion to dielectrics and inhibiting diffusion into silicon or dielectrics, and provide capability of electrodeposition of Cu [7]. From the aforementioned applications of metallic coating and thin films, it is realized that their fabrication often relies on non-equilibrium routes, such as a variety of ultrahigh vacuum techniques and electrodeposition/electroplating. The energetic adatoms landing on the substrate often first form nanometric epitaxial zone and then become 3D clusters during the growth process [8]. Moreover, the sophisticated compositions or the interaction between matrix atoms and impurity atoms as dopants and alloying elements often exert pinning effects. These factors result in the formation of abundant GBs among the columnar nanograins. Many thin films have been grown homo- or heteroepitaxially on single crystal templates or locally on polycrystalline templates, giving rise to the preferential texture in the films. Both the GB directionality and the preferential texture in the coatings or thin films should lead to mechanical and crystal anisotropies which greatly affect their performance in practical applications and hint at the property optimization along each direction.

1.2 Basics of anisotropy

In FCC single crystals, the Schmid factors mainly explains the crystal anisotropy. For a given crystal, different loading directions result in different sets of Schmid factors on the 24 slip systems FCC structure intrinsically has, eventually tailoring dislocations on different slip systems. Along with the dislocation populations, the dislocations would self-organize into certain low energy substructures with cellular shape. For instance, [111]-loading generally leads to planar-shaped cell substructures, whereas [100] direction renders spherical-shaped ones. To be specific, the Schmid factors of the four slip planes for three different loading directions are present in **Table 1**. Z. Q. Wang et al. introduced an H-factor based on Schmid factors to comprehend the deformation heterogeneity and the formula is expressed as [2].

$$H = \frac{\tilde{m}_{i,max} - \tilde{m}_{i,min}}{\sum_i \tilde{m}_i} \tag{3}$$

where $\tilde{m}_i$ is the sum of all three values of the Schmid factors for each slip plane under one loading orientation, and $\tilde{m}_{i,max}$ and $\tilde{m}_{i,min}$ are the maximum and minimum of $\tilde{m}_i$ out of four different planes, respectively. This causes the $[\bar{2}11]$-, [111]- and [100]-loadings to render the respective H-factor of 0.376, 0.333 and 0. Moreover, this suggests that $[\bar{2}11]$-loading brings about highest flow stress and most heterogeneous cellular structure, where [100]-loading the lowest flow stress and most homogeneous structure out of three loading conditions.

For polycrystalline cubic metallic materials, the Yield strength is associated to a Taylor factor, M, based on the critical resolved shear stress (CRSS) and the Taylor factor relies on the active slip systems at the grain level of a polycrystalline metal with crystallographic preferential texture in the aggregate. According to the classic dislocation mechanics [9], the Yield strength can be defined as

$$\Delta\sigma_y = M\left[\tau_0 + \left(\frac{\tau^* \mu b}{\pi(1-v)L}\right)^2\right] \tag{4}$$

where τ_0 and L are the lattice friction and the mean spacing between a dislocation source to obstacle, respectively. τ^* denotes the barrier shear stress for a single dislocation transmission across a GB and v, b and μ are the Poisson's ratio, Burgers vector and the shear modulus, respectively. The Taylor factor is intimately associated with the stacking fault energy (SFE) as well as the ratio, ξ, of CRSS for twinning and slip for FCC metals. Besides FCC structure, various crystalline materials are anisotropic. For example, various carbides hexagonal M_7C_3 (M = Fe, Cr, W, Mo) exhibited anisotropy and difference in chemical bonding along different crystallographic orientation determined the anisotropy and their elastic anisotropy could be further tailored by the different atomic arrangement controlled by multialloying [10]. Despite different mechanisms, some liquid crystals can be defined as the anisotropic fluids whose form is between the isotropic liquid and the anisotropic crystalline phase [11].

	[100]-loading	**[111]-loading**	**$[\bar{2}11]$-loading**
(111)	0.41, 0.41, 0.0	0.0, 0.0, 0.0	0.0, 0.0, 0.0
$(1\bar{1}1)$	0.41, 0.41, 0.0	0.27, 0.27, 0.0	0.41, 0.27, 0.14
$(\bar{1}\bar{1}1)$	0.41, 0.41, 0.0	0.27, 0.27, 0.0	0.41, 0.27, 0.14
$(\bar{1}11)$	0.41, 0.41, 0.0	0.27, 0.27, 0.0	0.27, 0.27, 0.0

Table 1.
Schmid factors of the four slip systems for three different loading directions. Reconstructed from reference [2].

1.3 Anisotropy in non-metallic 2D nanomaterials

The 2D non-metallic materials display remarked structural anisotropy due to the large interlayer spacing and comparably low interlayer cohesion and interaction, which causes that a monolayer of graphene could be readily mechanically exfoliated and hexagonal MoS_2 (h-MoS_2) with lamellar structure can be used as a solid lubricant owing to its superlubricity causing the facile glide among MoS_2 nanosheets. However, in the in-plane direction, the assumption of mechanical isotropy in 2D materials is premature just based on the six-fold symmetry in their hexagonal lattice when the isotropy has been assumed for some estimations of the elastic behaviors in carbon nanotubes. Prior researches unveiled that friction force exerted on both graphene and MoS_2 along in-plane 'zigzag' and 'armchair' directions of the hexagonal lattice gave rise to different results and friction tests along armchair direction resulted in larger friction forces. M. Dienwiebel et al. found the angular interval between two friction peak force being approximate 60° upon friction tests on graphite [12]. This suggests that the 2D materials with hexagonal lattice manifest a sixfold anisotropy with a 60° periodicity. Meanwhile, studies showed that the anisotropy in both graphene and MoS_2 has a thickness dependence [13]. 2D non-metallic nanomaterials have been often used as building blocks or components for micro/nano-electromechanical systems (M/NEMSs) and nanoelectronics. The anisotropy of those 2D materials have great influence on not only mechanical properties but also functional properties.

1.4 Anisotropy in metallic 2D thin films

Metallic coating and thin films have been largely fabricated adopting non-equilibrium ultrahigh vacuum techniques and electrodeposition. When the nuclei heterogeneously grow and then 3D clusters collide amid the coalescence process, forming intercrystalline interface. This process generally gives rise to nanocolumnar grains whose grain size is small, even in monolithic metals, in contrast to other equilibrium processes. **Figure 1** shows the structure zone diagram after energetic deposition of a thin film on a substrate, indicating that columnar grains preferentially being generated at different generalized temperature T/T_m and argon pressure [14]. The columnar structure could even exist in amorphous Al-Cr thin films prepared by the sputtering technique as a result of chemical segregation [15]. These 2D

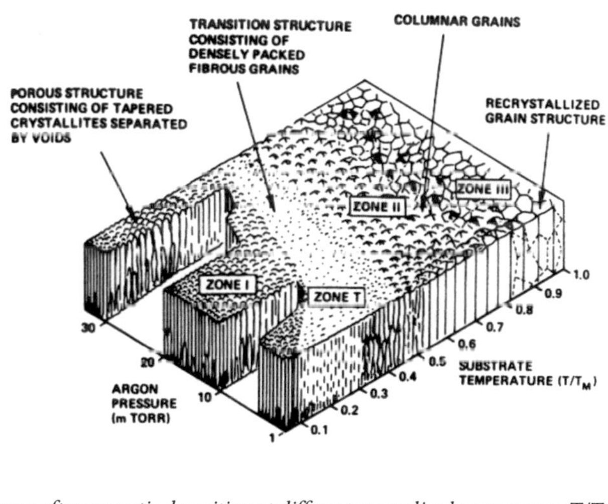

Figure 1.
Structure zone diagram after energetic deposition at different generalized temperature T/T_m *and argon pressure. Reprinted with permission from reference [14].*

metal coatings or thin films have higher hardness and strength, abiding by the well-known Hall–Petch relationship. However, the abundant columnar GBs with directionality are often the sites where the voids reside. The sluggish adatom kinetics and the shadowing effect from the surface e roughness lead to void formation residing at the columnar GBs and void-free GBs have lower cohesive energy when compared to the grain interiors. Z. S. You et al. found that nanotwinned (NT) Cu with columnar grains packed with horizontal coherent twin boundaries (TBs) experienced inhomogeneous deformation and columnar GBs were subjected to much larger plastic strain, compared to the grain interiors [16]. This caused one ambiguous puzzle, that is, the constant in the Tabor equation expressed as $H = C\sigma$ which translates the indentation hardness to the tensile strength often remarkably fails to fall in the proper proportionality range [9]. The proportionality constant, C, is dependent on the deformation mode under indentation and it had been empirically determined that $H/\sigma \approx 2.7$ for materials with high strain hardening coefficient and yield strength (elastic–plastic transition mode). This indicates the 2D metallic thin films with columnar GBs possess substantial structural anisotropy, despite the crystal anisotropy governed by either Schmid factor or Taylor factor [2, 8]. Metallic coatings or thin films have been used as protective, reflective, conductive components on apparatuses and devices. Comprehension toward the anisotropy of 2D metallic materials would substantially help improve their reliability and realize property optimization.

2. Fraction anisotropy in graphene

It has been known that graphene, graphene oxide and their composites exhibit mechanical anisotropy due to their characteristic of 2D extension [17, 18]. The investigations on the superlubricity of 2D nanomaterials have been also extensively conducted. A classic example is that M. Dienwiebel et al. studied the energy dissipation of a graphite at selective sliding directions on a Tribolever setup equipped with a tungsten tip and found the ultralow friction with the incommensurability nature [12]. Another example is M. Poot and H. S. J. Van der Zant adopted atomic force microscope (AFM) to measure force-distance relations on few-layer graphene and graphite flakes and discovered that a principle direction represents a stiffer direction than the others [19]. In contrast to those studies, a molecular dynamics simulation (MD) study is particularly selected to exhibit the anisotropic mechanical behaviors of graphene monolayers under uniaxial tensile condition along the zigzag and armchair directions [11]. 4.15×4.15 nm^2 square-shaped graphene monolayers with a thickness of 0.335 nm were fixed at one end and the tensile tests along the zigzag and armchair directions are present in **Figure 2a** and **b**. The relations between applied force and one unit cell are also present. The specific parameters of the non-equilibrium MD simulations can be found in the literature.

Regardless of the fracture patterns, the MD experiments first calculated the fracture stresses along the zigzag and armchair directions, which are 0.18 TPa at a strain of 32.48% and 0.21 TPa at 43.85%, respectively. In is worth noting that the predicted critical stresses and strains are anticipated to be higher than the empirical ones due to the idealism in the conditions of MD simulations. Prior to the crack formation, two test modes share similarity, that is, in the elastic region, the graphene monolayers regardless of testing directions could sustain large elastic deformation and upon crack formations, the crack propagated rapidly and led to the final fracture within 0.01% strain, suggestive of a brittle cleavage fracture. For the zigzag direction, within the strain from 32.484% to 32.489%, the crack propagated from one edge to the other edge, forming a zigzag-like fracture topography and the

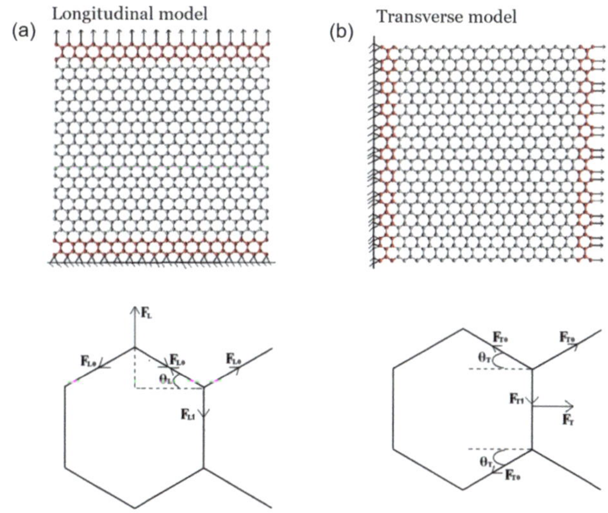

Figure 2.
Molecular dynamics simulation of tensile tests on 4.15 × 4.15 nm² square-shaped graphene monolayer along (a) zigzag and (b) armchair directions and the relations between applied force and one unit cell are present. Reprinted with permission from reference [20].

topological defects, whereas a rather smooth fracture feature was monitored as the strain varied from 43.859% to 43.866% under the test along armchair direction and the process left limited topological defects. It should be noted the five significant digits might be trivial in the real experiments but it was non-trivial in the MD simulations to capture detailed fracture process. The fracture evolutions along two directions were captured using snapshots in **Figure 3**. Since the C-C bonds have a critical strength, i.e. σ_{C-C}, it is anticipated that the direction of the applied force with respect to the hexagonal honeycomb lattice eventually governed the fracture mode and the analysis on the evolution of bond angles during the straining under two testing conditions is essential to decipher the different fracture mechanisms. Along the ziazag direction in **Figure 3a**, two 120° bond angles that evolved in a symmetrical pattern with the increase in the strain declined down to <90 ° and sustained substantial external strain, while the bonds in parallel to the tensile

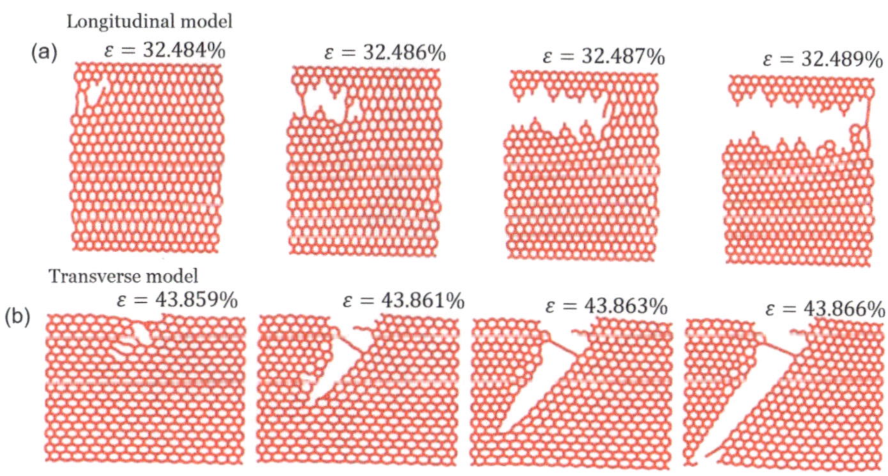

Figure 3.
Tensile strain-induced fracture process (a) along the zigzag direction and (b) along the armchair direction at various strain levels. Reprinted with permission from reference [20].

direction were elastically deformed until their critical strength was reached and then they broke, forming the broken hanging chains. In the other hand, the deformation along the armchair direction in **Figure 3b** caused the four 120° bond angles with a $\pm$ 30 ° relation with the applied force to decrease, transferring the hexagonal lattice into the quasi-rectangular shape, until any of the bonds except the two bonds normal to the testing direction broke. In the armchair mode, once triggered, the crack front would lead to a sequential instantaneous bond-breaking along the same direction and this left a 60° rupture along the armchair direction. Therefore, the two bonds parallel to the zigzag testing direction and the four bonds perpendicular to the armchair testing directions underwent the larger stress. As a result, the critical stress along the armchair is calculated to be $\sqrt{3}\sigma_{C-C}$, 1.73 times of the σ_{C-C} of the zigzag direction. In reality, the 0.21 TPa of the armchair direction is approximately 1.2 times of the 0.18 TPa of the ziazag direction, which is attributed to the evolving geometrical changes of the hexagonal structures amid the elongation processes. In addition, the geometrical changes of the hexagonal units during the straining determined the critical fracture stains for two testing directions. In order to validate the observation, the simulated dimensions were investigated to verify if the mechanical anisotropy has a size effect and it was found that the size effect was negligible.

3. Friction anisotropy in MoS2

The MoS_2 has similar lamellar structure as graphene and has been considered promising in the field of nanotribology, despites its various applications due to its functional properties. The inherent crystallographic characteristics of h-MoS_2 equips it with friction anisotropy pertaining to the effect of the lateral sliding direction on the friction behaviors or the commensurability/incommensurability conditions between two sliding planes. In the case of incommensurability where the two sliding surfaces have crystallographic nonmatching, ultralow friction is obtained and superlubricity occurs, which has been observed in 2D materials, such as MoS_2, graphene and highly oriented pyrolytic graphite to name a few. The superlubricity is related to the structural anisotropy. **Figure 4** presents the debris of five-layer thick MoS_2 after a wear test and the high resolution transition electron

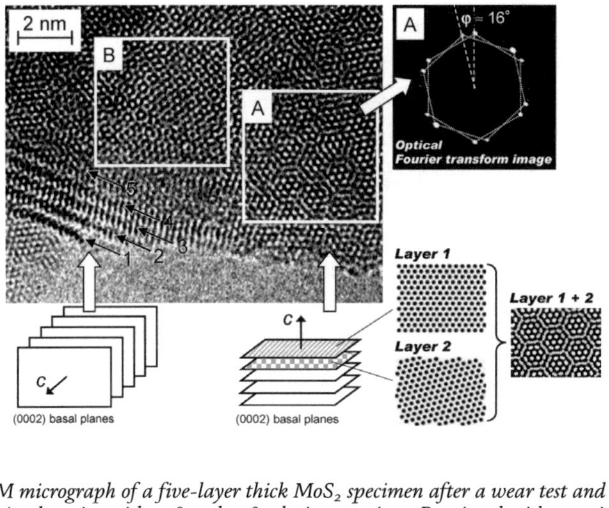

Figure 4.
High resolution TEM micrograph of a five-layer thick MoS_2 specimen after a wear test and the slide led to different mosaic lattice domains with 15° and 30° relative rotations. Reprinted with permission from reference [21].

microscopic (HRTEM) image shows different mosaic lattice domains as a result of 15° and 30° relative rotations between MoS_2 nanosheets [21]. Commonly, six-fold and two-fold symmetry of the friction behaviors have been captured on empirical and computational researches. Some two-fold symmetry of friction behaviors, namely 180° periodicity, have been attributed to the oriented linear wrinkles induced by the elastic deformation of the substrate and the testing conditions, one of which is the direction-dependent friction measured by an AFM tip with rotation. It was hypothesized that the tip rotation generated a variety of possible combination of the tip-specimen interfaces and the friction results might be able to reflect the genuine crystallographic pattern of the tested materials.

A study involving experimental and MD simulation results on the friction property of MoS_2 was present [22]. The direction-dependent friction behaviors were measured by changing the scanning direction and a 5 nm travel distance was applied to preclude the influence from the nanowrinkles. The atomic configuration and the scanning direction with respect to the lattice are illustrated in **Figure 5a**. **Figure 5b** presents the two friction loops consisting of forward and backward lateral scans, measured by AFM along zigzag and armchair directions, and it shows that the energy dissipated in each scan cycle of the tests along the armchair direction was 11 times higher than that of the tests along the zigzag direction. **Figure 5c** shows comparable simulation results and the quantitative discrepancy between the empirical and simulation results originates from the difference in tip conditions and the magnitude of the scanning speed and force. The Prandtl-Tomlinson model alleged that the friction at the atomic level relies on the height of the surface energy barrier and longer scanning length along the armchair direction would result in accumulated energy dissipation in comparison with the zigzag direction. Hence, the direction-dependent friction behaviors were examined using potential energy surface (PES). **Figure 6a** reveals a six-fold symmetry of the friction force in nN in comparison with the two-fold symmetry. To further comprehend the friction symmetry, PEC at various angular positions was observed with a 10° interval. **Figure 6b** reflects the cross-section energy profiles for the scans at 0°, 10° and 50°. **Figure 6c-j** show that PES possessed a 60° periodicity, e.g. the energy surfaces of the 0° and 60° being identical. Therefore, a friction anisotropy was explored at an atomic level, proving that the testing direction and tip-specimen contact quality greatly play significant roles in changing the energy landscape and affecting the friction behaviors. X. Cao et al. have exhibited that the friction behaviors of MoS_2 had a thickness effect [13]. In brief, the decrease in MoS_2 thickness down to a few nanometers could progressively weaken the anisotropy phenomenon and be more governed by the puckering effect.

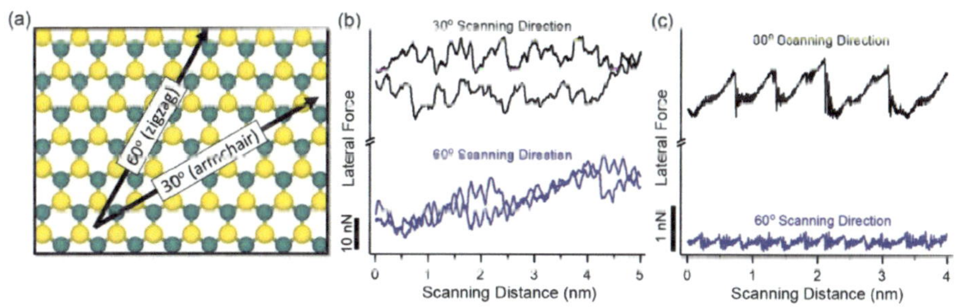

Figure 5.
(a) Atomic configuration of a MoS_2 monolayer in the simulations, indicating the armchair (30°) and zigzag directions (60°). (b) the experimental friction loops consisting of the forward and backward scanning along the armchair and zigzag directions. (c) the friction traces, due to tip-specimen contacts, predicted by the MD simulations along the armchair and zigzag directions. Reprinted with permission from reference [22].

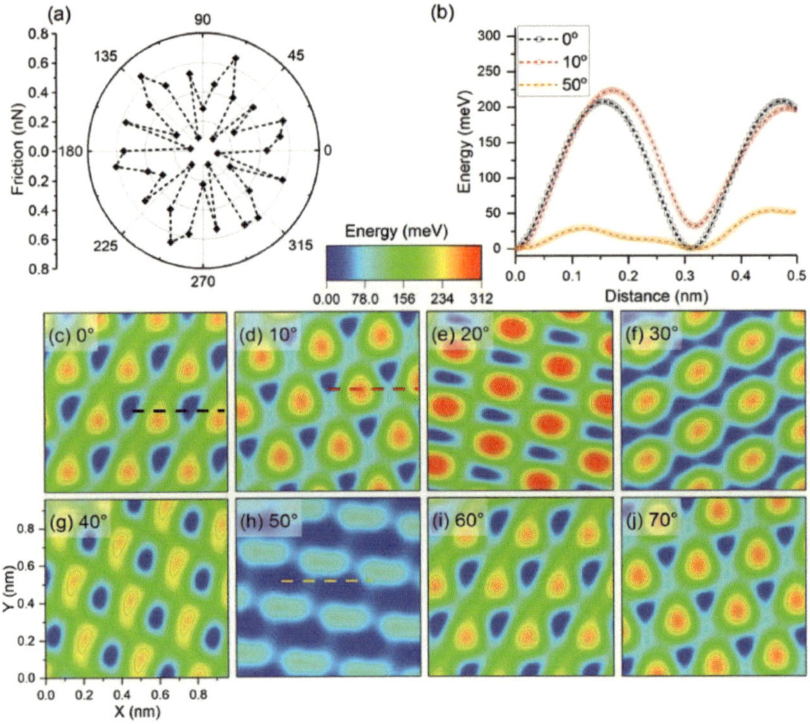

Figure 6.
(a) MD simulation that monitored a 6-fold symmetry in the results of friction tests on MoS_2 as a function of the rotational. (b) Cross-sectional barrier profiles along the scanning distance at a sample rotation angle of 0°, 10° and 50°. (c – j) potential energy surface (PES) calculated for the specimen rotated from 0° to 70°. Reprinted with permission from reference [22].

4. Mechanical anisotropy in Al-Fe thin films

Since the early 1950's when Hall and Petch empirically demonstrated that the Yield strength of metallic materials is inversely proportional to the square root of the microstructural features, researchers have put enormous efforts in refining the microstructure and thus developed ultrafine grained materials and nanocrystalline materials in order to lift mechanical strength for both fundamental exploration and practical applications. Non-equilibrium routes have been commonly used to shrink the grain size of the metallic materials and most of techniques, such as ultrahigh vacuum techniques and electrodeposition, produce 2D metallic materials, i.e. coatings and thin films. Researchers have found that the tensile strength collected from the tensile tests on thin film metals, especially alloys, fell short of the predicted strength translated from nanoindentation measurements according to the Taylor relation, i.e. $H = C\sigma$ where C is the proportionality constant. $H/\sigma = 3$ is often observed for materials with low strain hardening coefficient and low yield strength (fully plastic contact mode), whereas $1.1 < H/\sigma < 3$ is applicable for materials with high strain hardening coefficient and high yield strength (elastic–plastic transition mode). Most of time, nanoindentation studies showed empirically that $H/\sigma \approx 2.7$ for thin film metals with high strength. The off-proportionality has been often attributed to the voids potentially residing at the columnar GBs in thin film materials. However, the void size is proportional to columnar grain size and when the grain size is at nanoscale, the shadowing effect that originates from the 3D cluster growth should be negligible to cause void formation. Therefore, the directionality of the columnar GBs that contributes to the structural anisotropy has been largely ignored. Li et al. selected Al-Fe alloys produced by magnetron sputtering to investigate the anisotropy and tension-compression asymmetry along both the film in-plane and out-of-plane directions by adopting comprehensive in-situ micro-compression and tension techniques [9].

Figure 7a shows the dark-field TEM image and HRTEM image, suggesting that the Al-Fe alloys have abundant vertical GBs, which were identified as incoherent twin boundaries (ITBs) with a diffused feature, and an average grain size of ~5 nm. It is expected that the tiny grain size would greatly suppress the dislocation accumulation process that takes place in the plastic deformation of single crystals or polycrystalline materials with large grain size, making the deformation or fracture events more dominantly influenced by the directionality of the GBs. **Figure 7b-d** illustrate the micro-tension and compression experiments along in-plane and out-of-plane directions and exhibit the microsized specimens awaiting the in-situ experiments.

The experiment results showed that the out-of-plane compression experiments gave rise to a ~ 2 GPa strength and exhibited extensive deformability attributed to the grain coarsening, whereas in-plane compressions yielded a ~ 1.6 GPa strength but an intergranular shear deformation along the GBs, leaving the formation of shear bands. This deformation mode was governed by the maximum resolved shear stress. In addition, out-of-plane tensile experiments gave a tensile strength of ~1.8 GPa, comparable to the 2 GPa compressive strength, and a fracture mode governed by the intragranular shear propagation which were substantially deflected by vertical GBs. The apparent global engineering strain cannot be equated to the

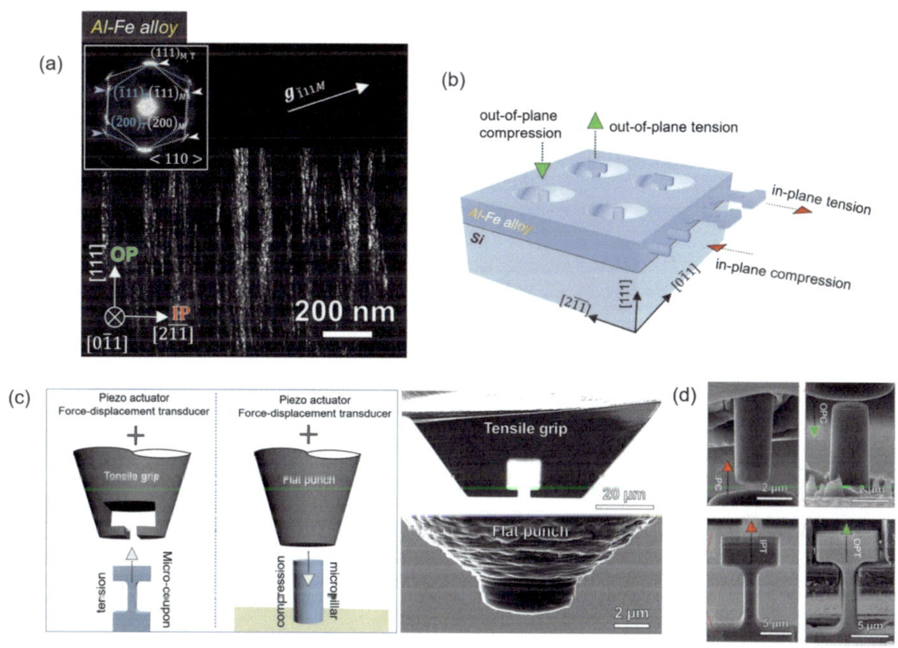

Figure 7.
(a) Cross-section dark-field TEM image showing abundant vertical GBs in a (111)-textured Al-Fe thin film and the neighboring columnar nanograins having twin relation. (b) Schematics illustrating how to perform micro-tension and compression experiments along both in-plane and out-of-plane directions. (c) Schematics of the setups of the in-situ *micro-tension and compression experiments along with the scanning electron microscope (SEM) images of the tensile gripper and the flat punch. (d) Micropillars and tensile micro-coupons awaiting the in-situ tension and compression experiments. Note that OPC, IPC, OPT and IPT denote out-of-plane compression, in-plane compression, out-of-plane tension, and in-plane tension, respectively. Reprinted with permission from reference [9].*

ductility of the common ductile materials with larger grain size and dislocation-dominated deformation mechanisms. In contrast, the in-plane tension experiments exhibited a relatively low strength of ~1.1 GPa and classic brittle behaviors governed by the nominal stress-induced fracture. The premature fracture propagated along GBs. It was found that the chemical combination of binary Al-Fe alloys did not satisfy the embrittlement criteria in the Gibson-Schuh model [23], meaning that the relation between applied tensile stress and the directionality of the vertical void-free GBs, i.e. vertical ITBs, mostly rendered the premature fracture phenomenon under in-plane tension mode rather than other factors including voids and GB embrittlement. **Figure 8** had summarized the major deformation or fracture mechanisms of the tension and compression tests along the in-plane and out-of-plane directions. It is noted that the anisotropy experienced in the Al-Fe thin films is different from the anisotropy in single crystals and polycrystals, governed by the Schmid factor and the Tylor factors. Moreover, the thin film alloys, including the Al-Fe, are also different from the isotropic nanocrystalline metals and alloys with textureless feature. However, dislocations were indeed captured in the differently deformed Al-Fe specimens. It was found that under compression, the ratio of the yield strength collected under out-of-plane compression mode and in-plane compression mode was ~1.25, which was mostly governed by the Taylor factors of two testing directions. Since the Al-Fe alloys have a strong (111) out-of-plane texture, the out-of-plane Taylor factor is 3.67. Moreover, the in-plane direction has no obvious texture or a weak (112) texture and the two possibilities rendered similar Taylor factor of 3.06. This two Taylor factors led to a strength ratio of 1.2, coinciding with the 1.25 collected experimentally. This indicates that the anisotropy in the Al-Fe thin films was both influenced by the directionality of the GBs and the

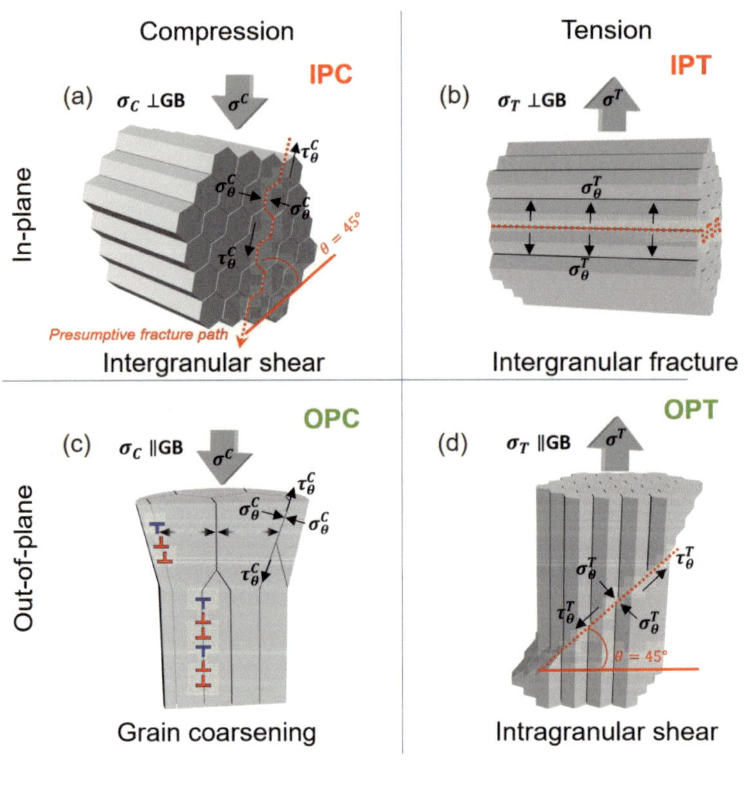

Figure 8.
Schematics that illustrate the deformation or fracture mechanisms for Al-Fe specimens tested at different modes along different directions. Reprinted with permission from reference [9].

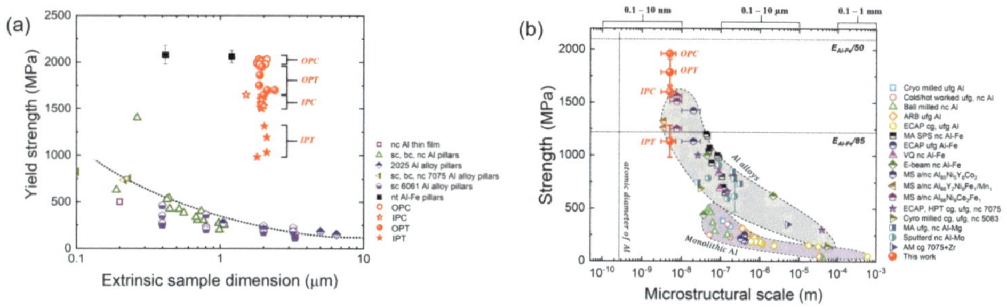

Figure 9.
(a) The yield strength or fracture strength, collected from different testing conditions, of Al-Fe thin films in comparison to the ones of other Al alloys with extrinsic dimensions at similar magnitudes. (b) the comparison of the strengths as a function of *intrinsic microstructural feature size effect between Al-Fe thin film and other pure Al and Al alloys. Reprinted with permission from reference [9].*

conventional Taylor factor. **Figure 9** plots the collected strengths under tension and compression along the in-plane and out-of-plane directions as a function of extrinsic specimen dimension and intrinsic microstructural feature size and it clearly manifested the anisotropy under both tension and compression modes. The Al-Fe alloys underwent negligible extrinsic size effect and are highly competitive as to the high strength.

Some thin films or coatings consisting of constituent elements with low stacking fault energy might have columnar grains packed with high-density horizontal coherent TBs (CTBs). Q. H. Lu et al. found that the dislocations were confined within the twin/matrix lamellae and the testing direction, the slip systems and the horizontal CTBs of the NT Cu could result in different dislocation structures and dislocation-CTB interactions, which rendered different hardening and softening modes and thus the anisotropy in metallic thin films made of constituent elements with low stacking fault energy [24]. Furthermore, it should be noted that not all the metallic thin films prepared by non-equilibrium methods possessed the conventional columnar GBs. Li et al. recently exhibited that manipulation of electrolytic solution with certain organic additive could potentially transfer the 3D cluster growth to a flat 2D layer-by-layer growth mode to facilitate the formation of TBs and suppress the formation of the columnar GBs from the island coalescence process [8]. The anisotropy of the NT metals mainly constructed by horizontal CTBs needs further investigation. Furthermore, the mechanical anisotropy in metallic materials could be also displayed from the dynamic strain-induced phase transformation. In a Transformation induced plasticity (TRIP) steel, a strong texture after rolling was obtained in the austenite and the texture in austenite gave rise to a higher martensitic transformation rate along the rolling direction, which contributed to a more pronounced TRIP effect and a higher strain-hardening rate [25, 26].

5. Conclusions

The synthesis, microstructural controls and the functional applications of 2D materials have been top trending research topics in the past 2 decades. However, the anisotropy of the 2D materials has not been put equal but actually exerts potent influence on not only their mechanical behaviors but also the multifunctional performance of materials and devices with 2D materials as components or building blocks. The unique microstructural characteristics of 2D materials result in distinct and intriguing structural and crystal anisotropy. As to the non-metallic 2D nanomaterials, such as graphene and MoS_2, the orientation of the applied stress with respect to the lattice often cause different interlayer friction, even the superlubricity, and the monolayer with the inherent crystallographic symmetry of the hexagonal honeycomb lattice also exhibited anisotropy when subjected to fracture. For the metallic thin films with 2D extension and limited thickness, the directional and abundant grain boundaries could influence the anisotropy comparison with the bulk single crystals or polycrystals whose anisotropy is primarily dominated by the Schmid factor or Taylor factor. It is anticipated that the sustainability and reliability of the materials and devices constructed by various 2D materials rely on the prominent anisotropy inside 2D materials. The in-depth comprehension toward the anisotropy of 2D materials would be also instructive to realize the orientation-dependent properties and the property optimization.

Author details

Qiang Li
School of Materials Engineering, Purdue University, West Lafayette, IN, United States

*Address all correspondence to: qnli2015@gmail.com

References

[1] Novoselov KS, Geim AK, Morozov SV, Jiang D, Zhang Y, Dubonos SV, Grigorieva IV, Firsov AA. Electric Field Effect in Atomically Thin Carbon Films. Science. 2004;306(5696): 666–669. DOI: 10.1126/science.1102896

[2] Wang ZQ, Beyerlein IJ, Lesar R. Plastic anisotropy in fcc single crystals in high rate deformation. Int. J. Plasticity. 2009;25:26–48. DOI: 10.1016/j.ijplas.2008.01.006

[3] Jian JY, Chang HL, Xu T. Structure and properties of single-layer MoS_2 for nano-photoelectric devices. Materials. 2019;12:198. ; Doi:10.3390/ma12020198

[4] Tan CL, Cao XH, Wu XJ, He QY, Yang J, Zhang X, Chen JZ, Zhao W, Han SK, Nam GH, Sindoro M, Zhang H. Recent Advances in Ultrathin Two-Dimensional Nanomaterials. Chem. Rev. (2017);117(9):6225–6331. DOI: 10.1021/acs.chemrev.6b00558

[5] Šiškins M, Lee M, Alijani F, Van Blankenstein MR, Davidovikj D, Van der Zant HSJ, Steeneken PG. Highly Anisotropic Mechanical and Optical Properties of 2D Layered As_2S_3 Membranes. ACS Nano. 2019;13(9): 10845–10851. DOI: 10.1021/acsnano.9b06161

[6] Moreno FP, Jakab MA, Tailleart N, Goldman M, Scully JR. Corrosion-resistant metallic coatings. Materials Today. 2008;11(10):14–23. DOI: 10.1016/S1369-7021(08)70203-7

[7] Hsiao HY, Liu CM, Lin HW, Liu TC, Lu CL, Huang YS, Chen C, Tu KN. Unidirectional Growth of Microbumps on (111)-Oriented and Nanotwinned Copper. Science. 2012;336(6084):1007–1010. DOI: 10.1126/science.1216511

[8] Li Q, Xue SC, Price P, Sun X, Ding J, Shang ZX, Fan Z, Wang H, Zhang Y, Chen YX, Wang HY, Hattar K, Zhang XH. Hierarchical nanotwins in single-crystal-like nickel with high strength and corrosion resistance produced via a hybrid technique. Nanoscale. 2020;12:1356–1365. DOI: 10.1039/C9NR07472D

[9] Li Q, Xue SC, Zhang YF, Sun X, Wang HY, Zhang XH. Plastic anisotropy and tension-compression asymmetry in nanotwinned Al–Fe alloys: An *in-situ* micromechanical investigation. Int. J. Plasticity. 2020;132:102760. DOI: 10.1016/j.ijplas.2020.102760

[10] Chong XY, Hu MY, Shan Q, Jiang YH, Li ZL, Feng J. Tailoring the anisotropic mechanical properties of hexagonal M7X3 (M=Fe, Cr, W, Mo; X= C, B) by multialloying. Acta Materialia. 2019;169:193–208. DOI: 10.1016/j.actamat.2019.03.015

[11] Asiya SI, Pal K, Kralj S, Thomas S. Nanomaterials dispersed liquid crystalline self-assembly of hybrid matrix application towards thermal sensor. Nanofabrication for Smart Na nosensor Applications. 2020:295–321. DOI: 10.1016/B978-0-12-820702-4.00013-1

[12] Dienwiebel M, Verhoeven GS, Pradeep N, Frenken JWM, Heimberg JA, Zandbergen HW. Superlubricity of graphite. Physical Review Letters. 2004; 92(12):12601. DOI: 10.1103/PhysRevLett.92.126101

[13] Cao XA, Gan XH, Lang HJ, Yu K, Ding SY, Peng YT, Yi WM. Anisotropic nanofriction on MoS_2 with different thicknesses. Tribology International. 2019;134:308–316. DOI: 10.1016/j.triboint.2019.02.010

[14] Thornton JA. Influence of apparatus geometry and deposition conditions on the structure and topography of thick sputtered coatings. Journal of Vacuum

Science and Technology. 1974;11:666–670. DOI: 10.1116/1.1312732

[15] Li Q, Shang ZX, Sun X, Fan CC, Su RZ, Richter NA, Fan Z, Zhang YF, Xue SC, Wang HY, Zhang XH. High-strength and tunable plasticity in sputtered Al–Cr alloys with multistage phase transformations. International Journal of Plasticity. 2021;137:102915. DOI: 10.1016/j.ijplas.2020.102915

[16] You ZS, Lu L, Lu K. Tensile behavior of columnar grained Cu with preferentially oriented nanoscale twins. Acta Materialia. 2011;59:6927–6937. DOI: 10.1016/j.actamat.2011.07.044

[17] Chu K, Wang F, Wang, XH, Huang DJ. Anisotropic mechanical properties of graphene/copper composites with aligned graphene. Materials Science and Engineering: A. 2018;713(24):269–277. DOI: 10.1016/j.msea.2017.12.080

[18] Jayan JS, Pal K, Saritha A, Seeraj BDS, Joseph K. Graphene oxide as multi-functional initiator and effective molecular reinforcement in PVP/epoxy composites. Journal of Molecular Struc ture. 2021;1230:129873. DOI: 10.1016/j.molstruc.2021.129873

[19] Poot M, Van der Zant HSJ. Nanomechanical properties of few-layer graphene membranes. Applied Physics Letters. 2008;92:063111. DOI: 10.1063/1.2857472

[20] Ni ZH, Bu H, Zou M, Yi H, Bi KD, Chen YF. Anisotropic mechanical properties of graphene sheets from molecular dynamics. Physica B. 2010;405(5):1301–1306. DOI: 10.1016/j.physb.2009.11.071

[21] Onodera T, Morita Y, Nagumo R, Miura R, Suzuki A, Tsuboi H, Hatakeyama N, Endou A, Takaba H, Dassenoy F, Minfray C, Pottuz LJ, Kubo M, Martin JM, Miyamoto A. A Computational Chemistry Study on Friction of h-MoS_2. Part II. Friction Anisotropy. J. Phys. Chem. B. 2010;114:15832–15838. DOI: 10.1021/jp1064775

[22] Vazirisereshk MR, Hasz K, Carpick RW, Martini A. Friction anisotropy of MoS_2: effects of tip-sample contact quality. J. Phys. Chem. Lett. 2020;11:6900–6906. DOI: 10.1021/acs.jpclett.0c01617

[23] Gibson MA, Schuh CA. Segregation-induced changes in grain boundary cohesion and embrittlement in binary alloys. Acta Materialia. 2015;95:145–155. DOI: 10.1016/j.actamat.2015.05.004

[24] Lu QH, You ZS, Huang XX, Hansen N, Lu L. Dependence of dislocation structure on orientation and slip systems in highly oriented nanotwinned Cu. Acta Mater. 2017;127:85–97. DOI: 10.1016/j.actamat.2017.01.016

[25] Chen SC, Huang CY, Wang YT, Huang CY, Yen HW.Role of the crystallographic texture in anisotropic mechanical properties of a newly-developed hot-rolled TRIP steel. Materials Science and Engineering: A. 2020;790:139683. DOI: 10.1016/j.msea.2020.139683

[26] Min KM, Jeong W, Hong SH, Lee CA, Cha PR, Han HN, Lee MG. Integrated crystal plasticity and phase field model for prediction of recrystallization texture and anisotropic mechanical properties of cold-rolled ultra-low carbon steels. International Journal of Plasticity. 2020;127:102644. DOI: 10.1016/j.ijplas.2019.102644

Section 2

Composites and Extraction Analysis

Chapter 5

Development of LDPE Crystallinity in LDPE/Cu Composites

Makki Abdelmouleh and Ilyes Jedidi

Abstract

This chapter summarizes the study of the filler (ie copper) effect on LDPE phasic composition in LDPE/Cu composites prepared in solution. During this research work, a particular effort is focused on the use of DSC under non-standard conditions. Therewith, the presence of copper microparticles has a great effect on the network phase than on the crystalline long-range-order phase of LDPE structure. Furthermore, LDPE phasic composition in absence and presence of copper microparticles is investigated by FTIR spectroscopy followed by a spectral simulation of the band that appeared at 720 cm^{-1} corresponding to the CH_2. Anywise, the two-phase model confirmed that no variation is observed of LDPE phase composition for all copper contents into LDPE/Cu films. However, with the three-phase model the orthorhombic phase fraction was found to be constant compared to the fraction of amorphous and that of network phase were found to increase and decrease respectively with increase in the copper particle load in the film. Overall, the thermal and structural behavior of LDPE in presence of copper particles allows this type to be used as phase change materials (PCMs) by adding a paraffin fraction in the LDPE/Cu composite. An update of the most relevant work carried out in the field of phasic characterization of polyethylene is presented in this chapter.

Keywords: LDPE/Cu composite films, LDPE crystallinity, LDPE Network phase, thermal stability

1. Introduction

Inorganic fillers perform an important role in the production of polymeric composites. Several value-added properties other than low cost, are gained through the use of fillers. Fillers can improve the mechanical [1–3] and thermal [4–7] properties, as well as optical and electrical properties [8–12] of polymeric materials composites. The polyethylene (PE) as one of the most widely used thermoplastics resins possesses excellent biocompatibility with human body and usually used as implantable material [13]. The PE/Cu composites have been developed in a large range of applications. The physical properties depend on the percentage of filler in the composites materials. The crystallinity of a semi-crystalline polymer (in particular polyethylene) mostly decides its physical and some times chemical characteristics. Thereby, it is important to understand the effects of metallic fillers on the phase composition of a semi-crystalline polymer matrix.

IntechOpen

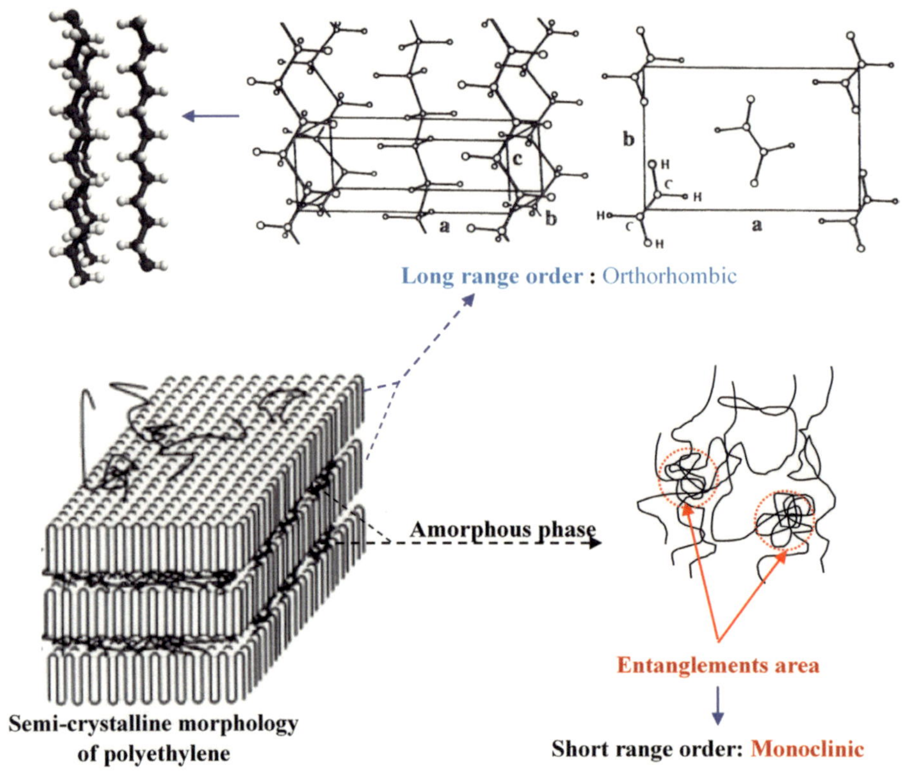

Figure 1.
Three-phase model. In areas of entanglement, a three-dimensional order at close range exists; this is the third phase.

The phase composition of polyethylene (PE) polymers has been described using a two phase or a three-phase model. In the two-phase model, the fraction of the long-range order phase is demonstrated by X-ray scattering and calorimetry but the other phase assumed to be liquid-like. Various techniques have been used to confirm the presence of a third-phase with mobility and order intermediate to that of the crystalline and amorphous phases for Polyethylene (PE). The name and the characteristics of this third phase depend on the type of crystal growth (bulk, solution, or fiber crystals) and on the technique used to analyze it. In NMR [14] and Raman [15–17] studies, the third phase in PE has been called interlamellar, inter-facial, or interzonal, in reference to its links to the crystalline and amorphous phases. The analysis of the melt is one of the techniques allowing suggesting the presence of more than two phases in a semi-crystalline polymer. Thus, ^{13}C NMR indicated the presence of two relaxation times [18, 19] relating to a heterogeneous fusion showing that the solid melts in a complex mode and incompletely.

The third phase, named the network phase, was identified following the characterization of the phase composition of different PE samples by slow Calorimetry, Calorimetry in non-standard conditions, FTIR and LCST technique [20–22]. This phase consists in a short range order located between the entanglements contained in the amorphous phase. In consequence, this network phase is heterogeneous with a range of order and tension (due to the entanglements) determined by the polymerization conditions [22], the sample history [21] and the charge load [23]. Taking this third phase into account, the phasic structure of polymers could be described as follows (**Figure 1**):

- Crystalline phase: the same as that described in the two-phase model, that is formed by sequences of ordered molecules in a crystal lattice forming a three-dimensional order at a long distance. The crystal unit cell is orthorhombic in the case of PE,

- Amorphous phase: disorderly and entangled chains,

- Network or interphase phase: formed by the entire network of entanglements contained in the amorphous phase. In these areas of entanglement, a certain three-dimensional crystallinity exists due to the organization of short sequences of chains in a short-distance crystal lattice. In the case of PE, the lattice of crystals of the small distance order would be of the orthorhombic and monoclinic type [24].

Using a three phase system (crystalline, network and amorphous) instead of a two phase system (crystalline and amorphous) to study the effect of filler particles on the crystallinity of LDPE can be more appropriate. This due to the existence of a third phase having a morphology intermediate between that associated with long-range order and that consisting of disordered chains. For a thin film of PE melted on a substrate composed by the zinc selenide (ZnSe) microparticles or the titanium dioxide (TiO2) nanoparticles, Bernazzani and Sanchez [24] found a relationship between the amorphous phase (of a two-phase system) and the Tm variation. With these results, they confirmed that the crystalline phase (long distance order) was not affected by the nature of the substrate or the particle size.

In the present chapter, we investigate the effect of the filler (ie copper) on LDPE phasic composition in LDPE/Cu composites prepared in solution. The main objective of this work is to highlight the effect of the presence of copper microparticles on the network phase and on the long and the short-range order crystalline phase of the LDPE structure. To determine the effect of the copper microparticles addition on the LDPE matrix crystallinity we used different physico-chemical characterization techniques have been used such as FTIR spectroscopy and Differential scanning calorimetry (DSC) at standard and non standard conditions (**Figure 2**).

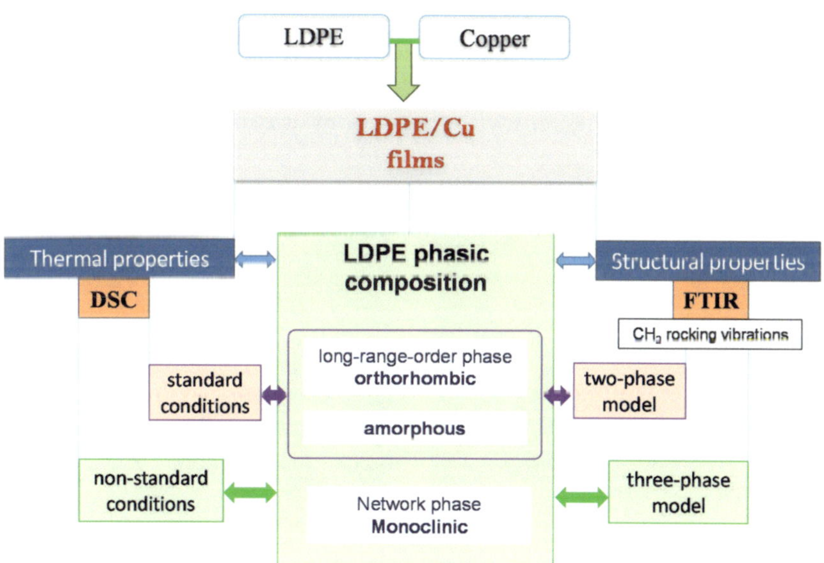

Figure 2.
Schematic representation of the thematic analysis of the chapter.

2. Preparation of the Cu/LDPE composite films

The low density polyethylene (LDPE) in pellets used in this work has a melting temperature of 107°C, a crystallinity around 40% (as determined by differential scanning calorimetry (DSC) with a heating rate of 10°C/min) and a density at room temperature of 0.9 g/cm^3. The particles sizes were < 38 μm and were supplied by Sigma-Aldrich. High purity toluene (99%) supplied by Sigma-Aldrish was used as solvent for the preparation of the solution made films. Different solutions of LDPE-toluene at 1% in the absence and in the presence of copper microparticles were prepared at 80° C. and cooled to room temperature. After total evaporation of the solvent in a Teflon mold, thin films (50 to 60 μm thick) were obtained [25].

3. Characterization of the Cu/LDPE composite films

3.1 Microscopy observation

Optical microscopy was used to investigate the distribution of the Cu microparticles in LDPE. The polarized optical microscopy photos of different LDPE/Cu composites were exposed in **Figure 3**. The copper particles distribution in the composite films are relatively uniform at both low (8%) and high (16%) copper contents.

3.2 Thermogravimetric analysis: TGA

The thermal stability of the composites was investigated by the Thermo Gravimetric Analysis (TGA). An amount of 5–10 mg of each sample were analyzed by Perkin-Elmer TGA7 Instrument from 35–600°C with a heating rate 20°C/min under nitrogen (N2) atmosphere.

The TGA curves for LDPE and its copper composites are shown in **Figure 4**. Neat LDPE showed a one-step decomposition process starting at 464°C due to the degradation of saturated carbon atoms in polyethylene that is displayed by peak on the curve DTG (**Figure 5**). The increase in the thermal stability of LDPE with increasing copper content showed in the composite films may be explained by the higher heat capacity (0.39 J/(Kg), compared to 0.18 J/(Kg) for PE) and thermal conductivity of Cu. This will result in the onset of the degradation of PE chains at higher temperatures. After the loss of this degradation, the level of mass loss are in good agreement with the amount of copper originally mixed into the samples.

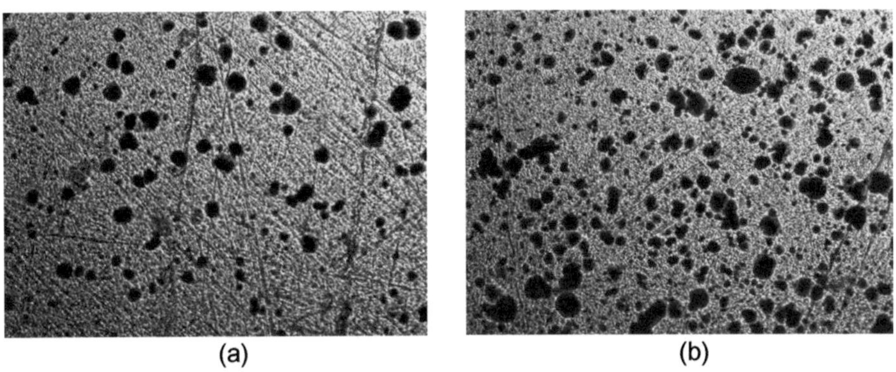

Figure 3.
Polarized optical microscopy photo of LDPE/Cu composites [100× magnification]: (a) 8%Cu; (b) 16% Cu.

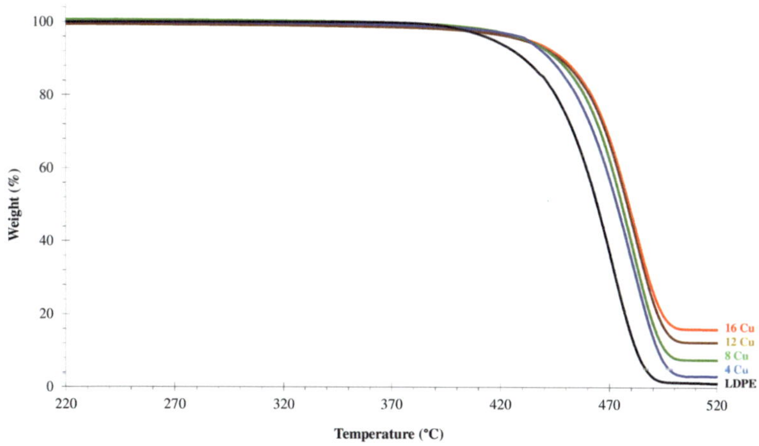

Figure 4.
TGA curves of LDPE and LDPE/Cu composites with different content of copper microparticles.

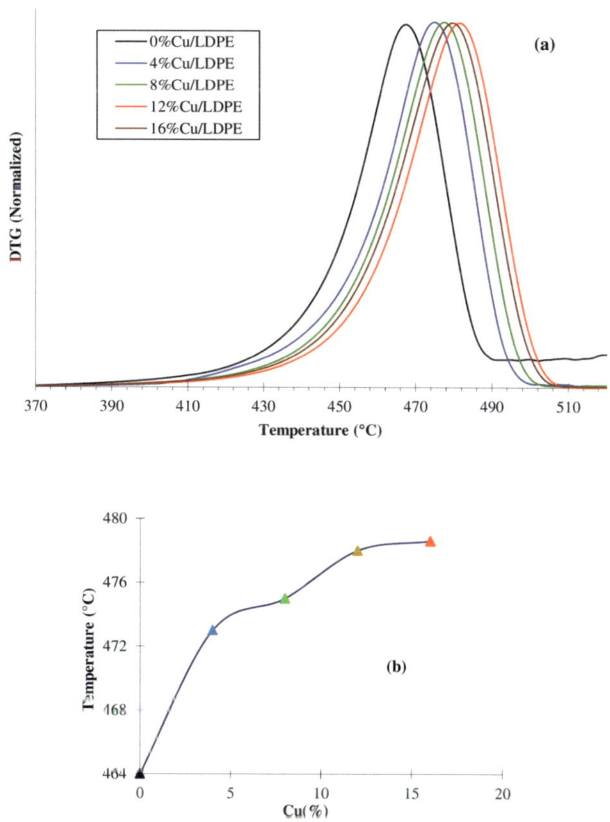

Figure 5.
(a) DTG for composite films at varying levels of copper. (b) Evolution of thermal decomposition temperature vs. Cu microparticles content.

DTG thermograms (**Figure 5**), thermal stability was clearly observed in LDPE/Cu composite films compared to unfilled LDPE film. This stability is reflected by a shift, towards high temperatures, of the decomposition peak of the composite films following the incorporation of copper microparticles into the LDPE The curves of the thermal decomposition temperature versus the amount of copper particles are presented in **Figure 5a**. This figure shows that the thermal decomposition temperatures of the LDPE/Cu are also higher than that of the pure LDPE, the tendency of the thermal decomposition temperature of the microcomposites increases with the increasing of the copper microparticles. They also show that the thermal decomposition temperature of the microcomposites reaches its peak as the amount of the copper microparticles is about 12 wt%, and then keeps on this scale when the amount of the copper microparticles is more than 12% wt in this experiment.

3.3 Differential scanning calorimetry: DSC

The DSC calorimeter is a Perkin-Elmer DSC7 with 20 mL/min flow of N_2. DSC in Standard Conditions (m = 2–3 mg and v = 10°C/min). The crystallization exothermic and endothermic curves of the neat LDPE and its composites with various copper micro-particles contents are illustrated in **Figure 6**. **Table 1** shows the T_m, T_C, the melting enthalpies ΔH_m, the crystallization enthalpies ΔH_C and the total crystallinity X_c values obtained for the neat LDPE film and the different LDPE/Cu composite films. The crystallinity X_c was determined as follows:

$$X_c = \frac{\Delta H_m}{W_m \Delta H_m^0} \tag{1}$$

Where ΔH°_m = 285 $J.g^{-1}$ is the heat of fusion for 100% crystalline PE and w_m is the weight fraction of polymeric matrix material in the composite [38].

Figure 6 show that both the melting temperatures range and the crystallization temperatures range decrease with increasing of copper micro-particles content. Data in **Table 1** indicate a slight decrease of the melting temperatures T_m but a slight increase of the crystallization temperatures T_C with the increasing of copper microparticles content. ΔH_m and ΔH_C decreases with the percentage of copper which is normal due to the decrease of the LDPE content in the composite film. The crystallinity Xc shows a slight decrease with the increase of the Cu content in the material showing

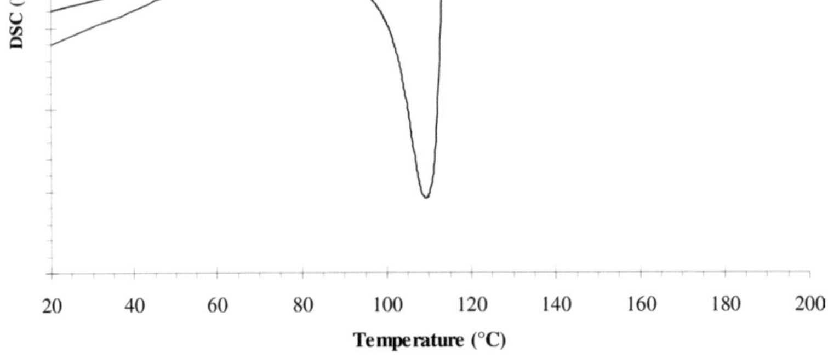

Figure 6.
Complete DSC in standard-condition heating and cooling scan of pure LDPE and LDPE/Cu composites.

LDPE/Cu (%)	100/0	96/4	92/8	88/12	84/16
ΔH_m (J/g)	112,3	108,9	94,4	95,4	89,8
T_m (°C)	110,7	109,3	110,1	109,6	109,7
X_c (%)	39,4	39,8	36,1	38,0	37,5

Table 1.
Thermal characteristics (T_m, T_C, ΔH_m, ΔH_C) and degree of crystallinity (X_c) of LDPE/Cu composites.

that the copper particles get a very limited effect on the orthorhombic long range order phase. This is confirmed by the fact that T_m does not did not suffer any noticeable variation. With DSC in standard conditions the strain created in the crystalline fraction particularly at T_m leaves an ordered fraction (network phase) un-melted [22].

DSC in non-standard conditions (m = 0.2–0.3 mg and v = 0.5°C/min): Due to the low values of mass and heating rate the sample (PE) will undergo a maximum heat flow [21]. The trace obtained for a neat LDPE film at the non-standard conditions (**Figure 7**) represents the endotherm of fusion of the orthorhombic crystals at 107.7°C (witch is the same T_m than that obtained by a fast T-ramp). On the other hand, at high temperatures the second endotherm obtained characterizes the melting of the strained short-range order crystals (network phase) [21, 22]. DSC in non-standard conditions (m = 0.2–0.3 mg and v = 0.5°C/min): Due to the low values of mass and heating rate the sample (PE) will undergo a maximum heat flow [21]. The trace obtained for a neat LDPE film at the non-standard conditions (**Figure 7**) represents the endotherm of fusion of the orthorhombic crystals at 107.7°C (witch is the same Tm than that obtained by a fast T-ramp). On the other hand, at high temperatures the second endotherm obtained characterizes the melting of the strained short-range order crystals (network phase) [21, 22].

Endotherms in non-standard conditions of neat LDPE and LDPE/Cu (84/16) composite (**Figure 7**) does not present any variation in the T_m value with the addition of copper particles. Never the less, the shape of the network melting endotherm of the different LDPE/Cu composite films looks very different when comparing it with that obtained for the neat LDPE film. In fact, in the traces of all the LDPE/Cu films, a more obvious separation between the Orthorhombic and the network

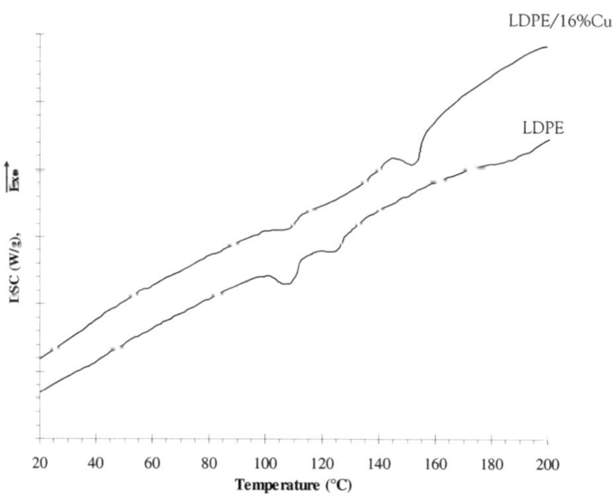

Figure 7.
Endotherm of neat LDPE at standard conditions (blue line) and endotherms in non-standard conditions of neat LDPE and LDPE/Cu(16%) composite (black line).

endotherms is observed. Also, more events happened in the melt expressed by the different melting/crystallization/melting observed in comparison to the more flat endotherm of the network melting endotherm of the neat LDPE film. These observations suggest that the presence of the micro-particles copper have more effect on the network phase than that can be observed on the crystalline long-range order phase. It is then more consistent to study the effect of the copper particles by analyzing the changes occurring on the network phase. The expansion during the temperature ramp operates a strain on the sample and the network phase is deformed. Over the melting of this phase showed by a succession of melting/crystallization/melting a complex phenomenon can observed [21]. In fact, the copper micro-particles participate in the strain applied on the sample witch explains the higher melting temperatures and the larger endotherms obtained for the network phase melting.

3.4 Infrared spectroscopy

The FTIR spectra of the studied films were obtained with a Perkin–Elmer Paragon 1000 FT-IR spectrometer used in transmission mode. The spectra's were treated using BOMEM GRAMS software for the determination of the study of the different phases of the LDPE.

3.4.1 Pure polyethylene thin films

The IR spectrum provided by the LDPE (**Figure 8**) shows absorption bonds characteristic of different vibration modes of methylene group. The main vibrations obtained on the spectrum are grouped together in **Table 2**. The elongation vibration, also called vibration of valence or "stretching", concerns the variation of the interatomic distance. When the molecule has symmetries, we can distinguish symmetrical or antisymmetric modes of elongation, which can be easily seen in the case of a methylene CH_2 group of LDPE (2919 cm^{-1} and 2845 cm^{-1}). In addition to stretching, the angles between the adjacent bonds of the CH_2 can vary, we then speak of deformation modes, which can be symmetric or asymmetric, and occur in the plane or out of the plane (1472 cm^{-1}, 1460 cm^{-1}, 720 cm^{-1} and 730 cm^{-1}). The presence of ($-CH_3$) is rarely desired when manufacturing a polymer. Indeed,

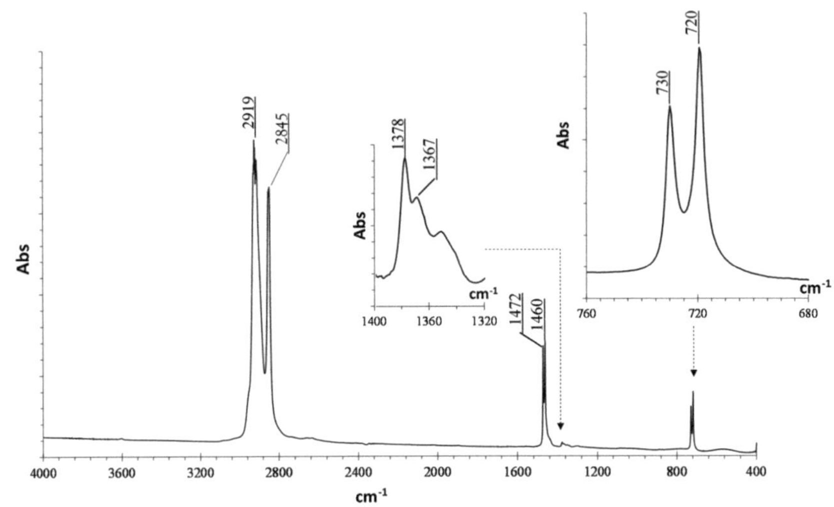

Figure 8.
FTIR spectra of LDPE film prepared from a solution.

Absorbance (cm^{-1})	Nature of vibrations		
720	γ C-H (amorphous phase)	Rocking CH_2	
730	γ C-H (cristalline phase)		
1460	δ C-H (amorphous phase)	Scissoring CH_2	
1472	δ C-H (cristalline phase)		
2845	ν CH_2 (s)	Symmetric stretching CH	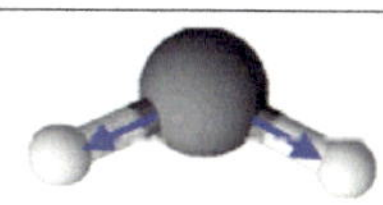
2919	ν CH_2 (as)	Asymmetric stretching CH	

Table 2.
Different absorption bands of LDPE.

a methyl group will decrease the density of the polymer by limiting the superposition of the layers, then making it of poorer quality. There is therefore an IR analysis method for determining the concentration of methyl group in a polymer as a function of the absorbance with two closely spaced bands (1378 cm^{-1} and 1368 cm^{-1}). These two bands visualized on the spectrum of the polyethylene used during this work (**Figure 8**) confirms the low density character of the polyethylene polymer and indicates its low rate of crystallinity by the existence of CH_3 groups in the polyethylene structure.

A series of standard infrared spectra as a function of temperature was first performed for pure LDPE thin film in order to understand the effect of interactions on the melting transition of LDPE. **Figure 9** shows the typical evolution of the infrared spectra of pure LDPE in the CH_2 rocking region as a function of temperature. The CH_2 rocking region is of particular interest. It shows a doublet attributed to the CH_2 rocking vibration. This doublet, caused by a field splitting, occurs only when the chains are in the orthorhombic crystalline phase. If the sample is not in the crystalline phase then only a broad peak around 725 cm^{-1} is observed. As the temperature approaches the melting temperature, the absorbance decreases until melting occurs and, as the polymer flows, the amount of sample effectively detected is reduced significantly leading to a sharp decrease in absorbance. The changes observed concern the relative intensity of the bands and their position.

3.4.2 Phase content by IR analysis: spectral simulations

An extensive research in the past have as objective the elucidation of the molecular structure of PE by IR. The assignment of the trans-trans and gauche conformations has led to the identification of ordered and less ordered regions. The ordered phases in the orthorhombic regions give a doublet in the rocking vibration at 730–720 cm^{-1} and bending vibration at 1500–1400 cm^{-1} due to the interaction between two chains in the unit cell. The analyzing the spectrum of molten PE and liquid, linear and cyclic alkanes gave mores information on mite less ordered region of a semi-crystalline polymer [26]. The amorphous regions have been associated with peaks at 725, 720, 1078 and 1300-1368 cm^{-1} [23].

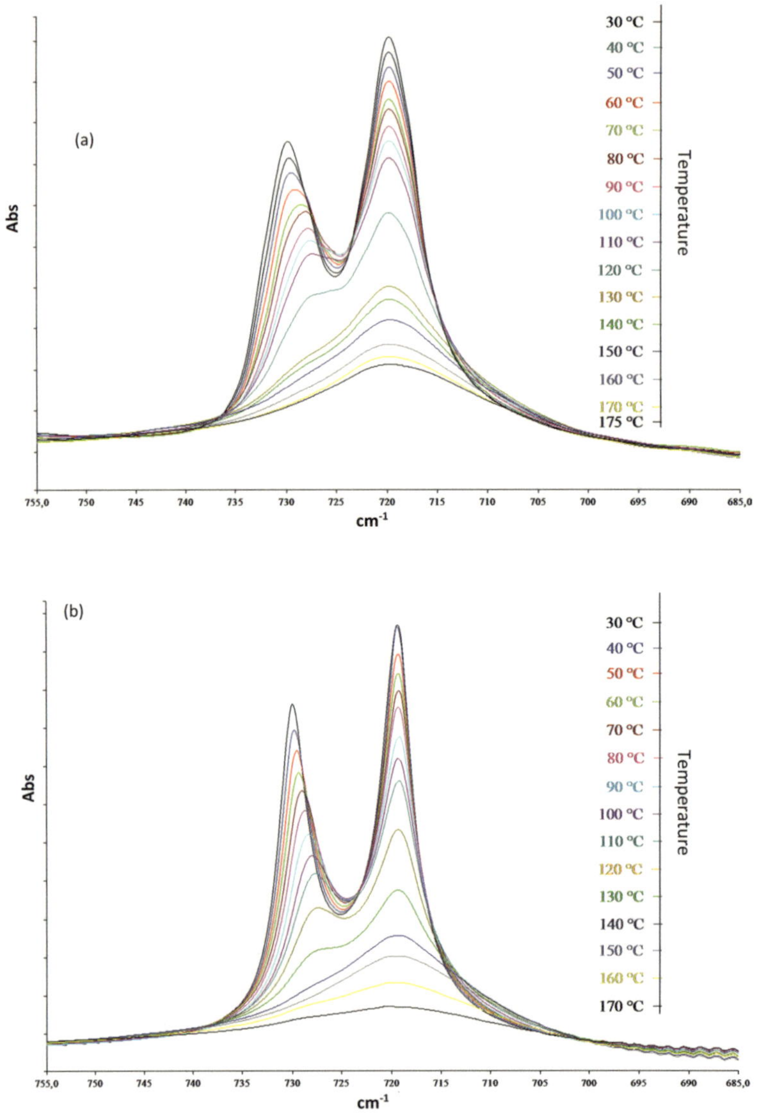

Figure 9.
Typical evolution of the CH_2 rocking region of the IR spectrum of a pure LDPE film (a) and LDPE/16%Cu, (b) composite film as a function of temperature.

The bands associations with a monoclinic-like organization have been studied on fibers and modified samples where the orthorhombic crystals had been deformed by shaking or drawing [27–34]. Thereby, the singlet at 715–718 cm^{-1} was replaces the doublet in the spectrum of these samples [35–37]. Using the spectral subtraction, these bands were also be detected in solution-grown crystals and on annealed melt-crystallized samples [35].

The superposition of the spectral region 670–770 cm^{-1} (**Figure 10a**) of LDPE and LDPE/Cu (84/16) shows a slight decrease in the peak around 730 cm^{-1} compared to that at 720 cm^{-1} in presence of micro-particles copper. Thus, to develop the crystallinity study of a LDPE/Cu composite film grown in solution, spectral simulation of the rocking CH_2 vibrations region for the different samples have been performed. Spectral simulations of the 770–670 cm^{-1} regions in the IR spectra were then performed by software BGRAMS/386 using a Gauss + Lorenz band shape with

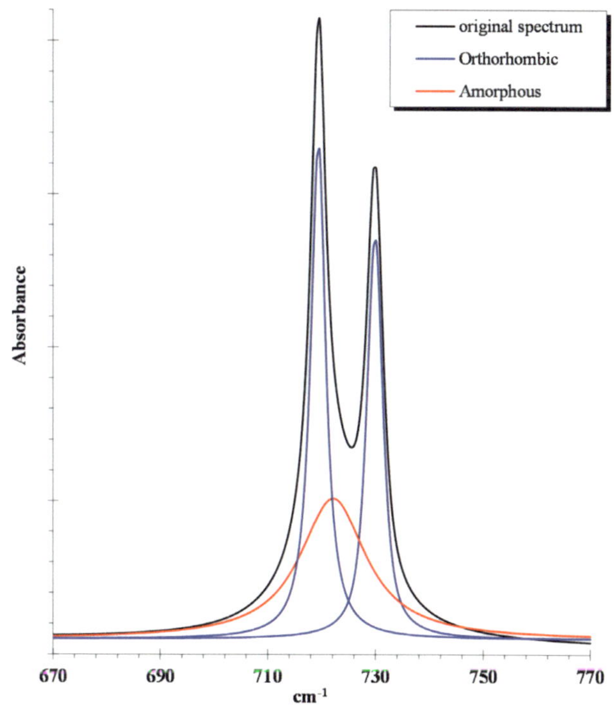

Figure 10.
(a) FTIR spectrum in the rocking region of LDPE and LDPE/Cu (84/16). (b) two-phase deconvolution of FTIR rocking region of LDPE and LDPE/Cu (84/16).

a band width at half-height of 3.5 cm^{-1} for the orthorhombic peaks, 10–16 cm^{-1} for the single-chain band or amorphous regions (725 cm^{-1}), and 15–19 cm^{-1} for thee monoclinic-like peak (717 cm^{-1}) [25]. The phase fractions were calculated from the simulated spectra using the equation:

$$\alpha_{phase(i)} = \frac{Integrated\ area_{phase(i)}}{Integrated\ \ area_{all\ the\ phases}} \tag{2}$$

The integrated absorbance of each band was rationed against the total area of the 770–670 cm^{-1} regions. The integrated absorbance of the two peaks at 730 and 720 cm^{-1}, both representing the orthorhombic fraction, were summed.

Using two phases model, the spectral simulation of transmission spectra of LDPE/Cu films with different copper content in the spectral range between 670 and 770 cm^{-1}, shows the presence of only three bands (**Figure 10b**):

- 725 cm^{-1} for amorphous phase
- 720 cm^{-1} and 730 cm^{-1} characteristic of the orthorhombic crystalline phase

Such deconvolution does not show any variation in the phase composition of the LDPE matrix for all copper contents (**Figure 11**). In fact, the fractions of orthorhombic and amorphous phases (respectively about 62% and 38%) are the same values obtained for the neat LDPE. Thus, copper does not have any noticeable effect on the structure of LDPE in the model of two phases [25].

Using three phases model, **Figures 12** and **13** shows four-band deconvolution for neat LDPE film (0% Cu) and a 16% copper loaded LDPE/Cu composite film. The orthorhombic peaks at 730 and 720 cm^{-1} appeared to be unchanged for all copper content witch comforts the thesis that such phase is not affected by the copper particles presence. Nevertheless, the peaks assigned to the amorphous phase (725 cm^{-1}) and to the network phase (717–715 cm^{-1}) had their integrated area respectively increased and decreased with the copper percentage in the film (**Figure 14**). This observation means that when three phases were introduced, the amount of the orthorhombic phase was found to be constant. However, starting at 4% copper content, the amount of the amorphous and that of the network phase were found to respectively increase and decrease with the increase of copper particles load in the film. This result are in agreement with the non-standard DSC observations mentioned

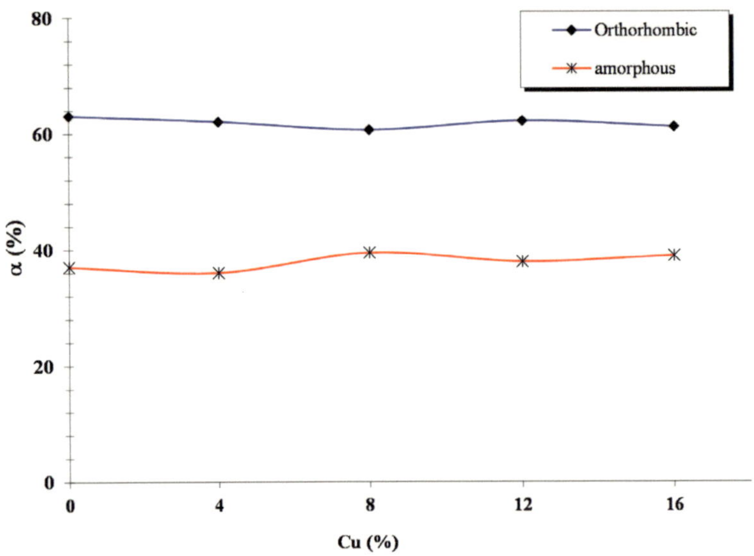

Figure 11.
Fractions of crystalline and amorphous phases of LDPE as functions of copper particle content in LDPE determined by FTIR using two-phase model simulation.

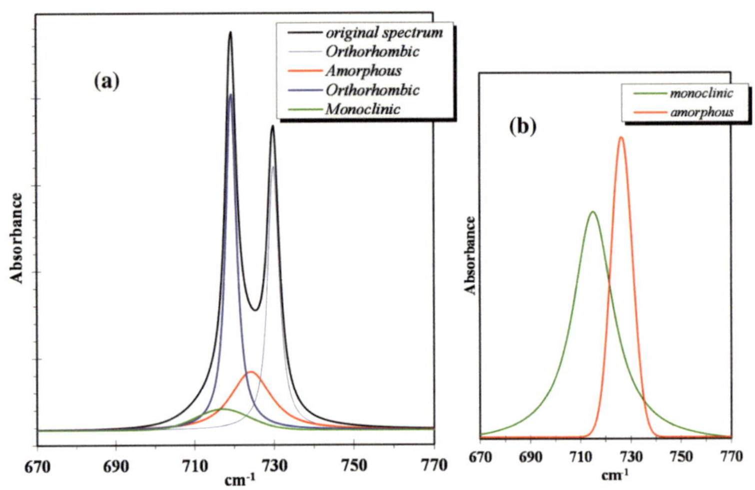

Figure 12.
FTIR spectrum in the rocking region showing three phases (orthorhombic, monoclinic, amorphous) of pure LDPE (a) Monoclinic and amorphous band evolution (b).

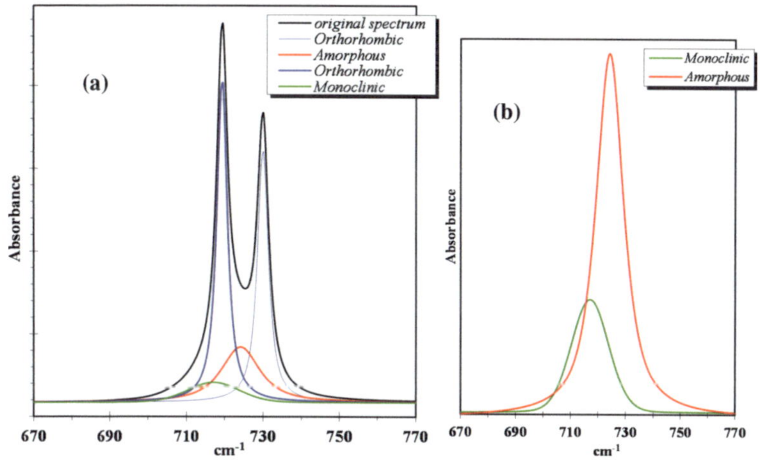

Figure 13.
FTIR spectrum in the rocking region showing three phases (orthorhombic, monoclinic, amorphous) of LDPE/16%Cu composite (a). Monoclinic and amorphous band evolution (b).

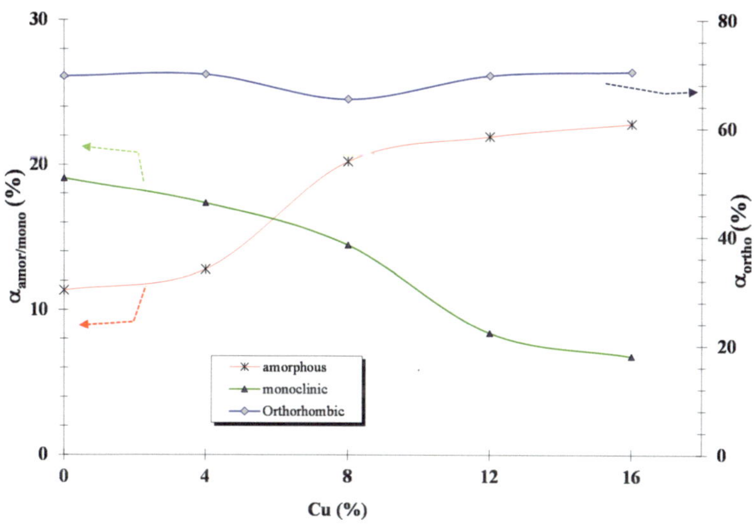

Figure 14.
Fractions of crystalline and amorphous phases of LDPE as functions of copper particle content in LDPE determined by FTIR using three-phase model simulation.

above and supporting the fact that the changes in the polymer phase composition due to the presence of the copper micro-particle is better seen when considering the existence of the network phase. The copper micro-particles lower probably the total crystallinity of the polymer matrix considering a three phase model, and do not show any effect when considering a two phase model [25].

The physical reason probably due to the fact that the copper particles may oppose to the formation of physical entanglements in their close environment. In consequence, the entanglements concentration as well as of the network phase content decrease with the increase of fraction copper and so does the network phase content. Other than the effect of copper micro-particles the polymer matrix seems to adopt the relaxed conformation of the amorphous phase due to the decrease of the entanglements concentration. The orthorhombic phase was not affected by the presence of copper micro-particles. However, with copper nanoparticles seen to have a clear effect on the long-range order of LDPE matrices in LDPE/Cu nanocomposites [38]. The difference might be due to the influence of the size effect of the copper nanoparticles.

4. Conclusion

LDPE/Cu composites prepared in solution was characterized by optic, spectroscopic and thermal analysis. The optical observation suggested that the copper powder particle distributions were found to be relatively uniform at both low and high copper contents. The presence of copper particles can improve the thermal stability of the composite since a maximum increment of 14°C is obtained comparing with the pure LDPE as shown on the TGA thermogram results. The results of DSC in standard conditions show that Cu content has little influence on the crystallinity X_c of LDPE. However, the trace of DSC at non-standard conditions suggested that the presence of copper microparticles has more effect on the network phase than that can be observed on the crystalline long range order phase. FTIR spectroscopy was used to study the phase content of LDPE in LDPE/Cu non-oriented composite films by investigate the CH_2 rocking vibrations. Spectral simulation of the transmission spectra performed using a tow phase model show that the copper contents does not any effect in the phase composition of the LDPE matrix. According of three phases model the amount of the orthorhombic phase was found to be constant. However, the amorphous and the network phase fraction were found to respectively increase and decrease with the increase of copper particles load in the film. The presence of an inorganic filler (Cu) in an organic polymer matrix (LDPE) forms a hybrid material that merges the properties of the two families of materials that compose it. This type of LDPE/Cu composite material which exhibited good structural and thermal stability as a function of the Cu fraction can be used as a phase change material (PCM) by the addition of a suitable oil phase.

Author details

Makki Abdelmouleh[1]* and Ilyes Jedidi[2]

1 Laboratoire de Sciences des Matériaux et de l'Environnement (LMSE), Université de Sfax, Tunisia

2 Engineering Department, College of Applied Science of SOHAR, SOHAR, Oman

*Address all correspondence to: makki.abdmouleh@yahoo.fr; makki.abdmouleh@fss.rnu.tn

References

[1] Bare W, Albano C, Reyes J, Dominguez N. Effect of irradiation on the mechanical properties of high-density polyethylene reinforced with metallic fibres. Surface and Coatings Technology. 2002;158:404-407. DOI: https://doi.org/10.1016/S0257-8972(02)00259-1

[2] Cristián S, Felipe B, Daniel C, Francesca AS, Alejandro C, Nicolás A, Lina MR, María TU, Pablo R, Andrés O, Tatiana G, Carlos L, Paula AZ. Mechanical and Antimicrobial Polyethylene Composites with CaO Nanoparticles. Polymers 2020, 12, 2132; doi:10.3390/polym12092132. DOI: https://doi.org/10.3390/polym12092132

[3] Anderson BC, Bloom PD, Baikerikar KG, Sheares VV, Mallapragada SK. Al-Cu-Fe quasicrystal/ultra-high molecular weight polyethylene composites as biomaterials for acetabular cup prosthetics. Biomater. 2002;23:1761. DOI: https://doi.org/10.1016/S0142-9612(01)00301-5

[4] Kuljanin J, Vuckovié M, Comor MI, Bibic N, Djokovic V, Nedeljkovic JM. Influence of CdS-filler on the thermal properties of polystyrene. Europ. Polym J. 2002;38:1659. DOI: https://doi.org/10.1016/S0014-3057(02)00043-5

[5] Weidenfeller B, Höfer M, Schilling FR. Thermal conductivity, thermal diffusivity, and specific heat capacity of particle filled polypropylene. Composites Part A: Appl. Sci. Manufac. 2004;35(4):423-429.

[6] Kim YD, Oh NL, Oh S-T, Moon I-H. Thermal conductivity of W–Cu composites at various temperatures. Mater Lett. 2001;51:420-424. DOI: 10.1016/S0167-577X(01)00330-5.

[7] Korab J, Stefanik P, Kavecky S, Sebo P, Korb G. Thermal conductivity of unidirectional copper matrix carbon fibre composites. Compo. Part A: Appl. Sci. Manufac. 2002;33(4):577-581. DOI: https://doi.org/10.1016/S1359-835X(02)00003-9

[8] Kaushik P, Asiya Si, Gharieb SS, M. Abd E, Rajesh K, Ahmed I, El-Batal, Samo K, Sabu T. Cutting edge development on graphene derivatives modified by liquid crystal and CdS/TiO_2 hybrid matrix: optoelectronics and biotechnological aspects. Critical Reviews in Solid State and Materials Sciences, DOI: 10.1080/10408436.2020.1805295

[9] Dang Z-M, Zhang Y-H, Tjong S-C. Dependence of dielectric behavior on the physical property of fillers in the polymer-matrix composites. Synthetic Metals. 2004;146(1):79-84. DOI: https://doi.org/10.1016/j.synthmet.2004.06.011

[10] Kaylon DM, Birinci E. Electrical Conductivity Of A Graphite Based Composite As Affected By The Degree Of Mixedness Of Graphite In The Elastomeric Matrix (364). Society of Plastics Engineers ANTEC Technical Papers. 2002;2:1716-1720.

[11] Mamunya YP, Zois H, Apekis L, Lebedev EV. Influence of pressure on the electrical conductivity of metal powders used as fillers in polymer composites. Powd. Tech. 2004;140:49-55. DOI: https://doi.org/10.1016/j.powtec.2003.11.010

[12] Das NC, Chaki TK, Khastgir D. Effect of processing parameters, applied pressure and temperature on the electrical resistivity of rubber-based conductive composites. Carbon. 2002;40(6):807-816. DOI: https://doi.org/10.1016/S0008-6223(01)00229-9

[13] Araya R, G´omez-Mora H, Vera R, Bastidas JM. Human spermatozoa motility analysis in a Ringer's solution

containing cupric ions. Contraception. 2003;67:161-163. https://doi.org/10.1016/S0010-7824(02)00477-8

[14] Kitamura R, Horii F, Murayarma K. Phase structure of lamellar crystalline polyethylene by solid-state high-resolution carbon-13 NMR detection of the crystalline-amorphous interphase. Macromol. 1986;19:636-643.

[15] R. Matter, Stille W, Strobl GR. Transition regions and surface melting in partially crystalline polyethylene: A raman spectroscopic study. J. Polym. Sci (B). Polym. Phys. 1993;31:99-105. DOI: https://doi.org/10.1002/polb.1993.090310113

[16] Strobl GR, Hagedorn W. Raman spectroscopic method for determining the crystallinity of polyethylene. J. Polym. Sci: Polym. Phys. 1978;16:1181-1193. DOI: https://doi.org/10.1002/pol.1978.180160704

[17] Glotin M, Mandelkern L. A Raman spectroscopic study of the morphological structure of the polyethylenes. Colloid Polymer Sci. 1982;260:182-192.

[18] Gennes P-G. Tight Knots. Macromol. 1984;17:703-704.

[19] T Bremner, Rudin A. Persistence of regions with high segment density in polyethylene melts. J. Polym. Sci (B). Polym. Phys. 1992;30:1247-1260. DOI: https://doi.org/10.1002/polb.1992.090301110

[20] Bernazzani P, Bich VT, Phuong-Nguyen H, Haine A, Chapados C, Dao Lé H, Delmas G. FTIR analysis of the phase content in low-density polyethylene. Can. J. Chem. 1995;76:1674-1687. DOI: https://doi.org/10.1139/v98-159

[21] Jedidi I, Ferhat-Hamida Z, Delmas G. Calorimetry in nonstandard conditions: The noncrystalline phases of linear polyethylene. J. Polym. Sci. Part B: Polym. Phys. 2007;45:1932-1949. DOI: https://doi.org/10.1002/polb.21207

[22] Delmas G. Irreversible formation of a network during melting/dissolution of nascent PE. J. Polym. Sci. Part B: Polym Phys. 1993;31:2011-2018. DOI: https://doi.org/10.1002/polb.1993.090311314

[23] Tiemblo P, Guzmán J, Serrano R, Hoyos M, García N. Evidence of a monoclinic-like amorphous phase in composites of LDPE with spherical, fibrous and laminar nanofillers as studied by infrared spectroscopy. Europ. Polym. J. 2009;45:30-39. DOI: https://doi.org/10.1016/j.eurpolymj.2008.09.040

[24] Bernazzania P, Sanchez RF. Effect of substrate interactions on the melting behavior of thin polyethylene films. Eur. Phys. J. E. 2008;26:427-434. DOI: https://doi.org/10.1140/epje/i2007-10344-7

[25] Abdelmouleh M, Jedidi I, Khitouni M, Ben Salah A, Kabadou A. LDPE phase composition in LDPE/Cu composites using thermal analysis and FTIR spectroscopy. J. Appl. Spect. 2011;78(2):191-199. DOI: https://doi.org/10.1007/s10812-011-9443-8

[26] Maroncelli M, Qi SP, Strauss LH, Snyder RG. Nonplanar Conformers and the Phase Behavior of Solid Alkanes. J. Am. Chem. Soc. 1982;804:6237-6247.

[27] Bank M, Krimm S. Mixed crystal infrared study of chain folding in crystalline polyethylene. J. Polym. Sci. Part A. 1969;27:1785-1809. DOI: https://doi.org/10.1002/pol.1969.160071014

[28] Krimm S. Infrared spectra of high polymers. Adv. Polym. Sel. 1960.2:51-172. DOI: https://doi.org/10.1007/BFb0050351

[29] Yoshikaru K, Krimm S. Infrared studies of the role of monoclinic

structure in the deformation of polyethylene. J. Macromol, Sci. Phys. B. 1970;4:461-472. DOI: https://doi.org/10.1080/00222347008229368

[30] Wedgewood AR, Seferis JC. Structural characterization of linear polyethylene by infrared spectroscopy. Pure. Chem. 1983;55(5):873-892.

[31] Agosti G, Zerbi G, Ward IM. Structure of the skin and core of ultradrawn polyethylene films by vibrational spectroscopy. Polymer. 1993.34:4219-4229. DOI: 10.1016/0032-3861(92)90261-T

[32] Zerbi G, Gallino G, Fanti N, Baini L. Structural depth profiling in polyethylene films by multiple internal reflection infra-red spectroscopy. Polymer. 1989;30:2324-2327. DOI: https://doi.org/10.1016/0032-3861(89)90269-3

[33] Painter PC, Havens J, Hart WW, Koenig JL. A fourier transform infrared spectroscopic investigation of polyethylene single crystals. II. Fine structure of the CH_2 rocking mode. J. Polym. Sci, Polym, Phys. 1977;15(7):1237-1247. DOI: https://doi.org/10.1002/pol.1977.180150709

[34] Siesler HW. Rheo-optical Fourier-Transform Infrared Spectroscopy (FTIRS) of polymers—6: Changes of crystal-axes orientation and state of order during uniaxial elongation of high-density polyethylene. Infrared Phys. 1984;24:239-244. DOI: https://doi.org/10.1016/0020-0891(84)90076-9

[35] Yoshikaru K, Krimm S. Infrared studies of the role of monoclinic structure in the deformation of polyethylene. J. Macromol, Sci. Phys. B. 1970;4:461-472. DOI: https://doi.org/10.1080/00222347008229368

[36] Xia X, Xie C, Cai S, Non-isothermal crystallization behavior of low-density polyethylene/copper nanocomposites. Thermoch act. 2005;427:129-135. DOI: https://doi.org/10.1016/j.tca.2004.09.002

[37] Xia X, Cai S, Xie C. Mater. Preparation, structure and thermal stability of Cu/LDPE nanocomposites. Chemist. Phys. 2006;95:122-129. DOI: https://doi.org/10.1016/j.matchemphys.2005.05.010

[38] Bryan Ellis, Polymers - a property database. Boca Raton (FL): CRC Press, 2000.

Chapter 6

Study of the Equilibrium of Nitric Acid with a Solution of TBP/IP6

Munoz Ayala Israel and Vera Roberto Carlos

Abstract

The behavior of the tri-n-butylphosphate (TBP) for a Liquid–liquid extraction (LLE) system is well known. To establish a new LLE system, the calculation of the equilibrium to establish an extraction system of TBP and inositol hexaphosphate (IP6) needs to be done. First, the change in the activity coefficient of TBP/IP6 related to the activity of water and TBP/IP6 concentration in the H2O–TBP/IP6–dodecane system, then the degradation of nitric acid in the system should be evaluated to assess the equilibrium. The proposed system consists of a solution of 30% of TBP and 10% of IP6 in nitric acid and dodecane. As main results, we discussed the value of the dissociation degree of nitric acid, the molar and volumetric fractions, the molar activity of the organic and aqueous phases and activities coefficients.

Keywords: equilibrium, TBP, IP6, extraction system

1. Introduction

Liquid–liquid extraction ion-exchange (LLE-IE), also known as solvent extraction and partitioning, is a method to separate compounds or metal complexes, based on their relative solubilities in two different immiscible liquids, usually water (polar) and an organic solvent (non-polar) [1]. There is a net transfer of one or more species from one liquid into another liquid phase, generally from aqueous to organic. The transfer is driven by chemical potential, i.e., once the transfer is complete, the overall system of chemical components that make up the solutes and the solvents are in a more stable configuration (lower free energy). The solvent that is enriched in solute(s) is called extract. The feed solution that is depleted in solute (s) is called the raffinate. This type of process is commonly performed after a chemical reaction as part of the work-up, often including an acidic work-up [2].

From a hydrometallurgical perspective, solvent extraction is exclusively used in separation and purification of uranium and plutonium, zirconium and hafnium, separation of cobalt and nickel separation, and purification of rare earth elements etc., its greatest advantage being its ability to selectively separate out even very similar metals. One obtains high-purity single metal streams on 'stripping' out the metal value from the 'loaded' organic wherein one can precipitate or deposit the metal value.

One of the well-known applications of a LLE in hydrometallurgical techniques is the PUREX (plutonium uranium redox extraction) which is a chemical method used to purify fuel for nuclear reactors or nuclear weapons. PUREX is the de facto

IntechOpen

standard aqueous nuclear reprocessing method for the recovery of uranium and plutonium from used nuclear fuel (spent nuclear fuel or irradiated nuclear fuel). It is based on liquid–liquid extraction ion-exchange [3].

It is not the intention of this research work to stablish a new PUREX methodology but to study the equilibrium of a LLE-IE based on TBP and IP6. The behavior of TBP and nitric acid (HNO_3) in the solvent extraction process has been studied, which has detected good stability, through laboratory tests, pilot tests and plant work.

IP6 is a unique natural substance found in plant seeds. It has received considerable attention due to its effects on mineral absorption. Impairs the absorption of iron, zinc and calcium and may promote mineral deficiencies. IP6 is a six-fold dihydrogenphosphate ester of inositol (specifically, of the myo isomer), also called inositol hexakisphosphate or inositol polyphosphate (IP6). At physiological pH, the phosphates are partially ionized, resulting in the phytate anion [4].

IP6 has had a high value for the nuclear industry, as it has studied as a complement to the recovery of uranium in seawater [3] and as a bio-recovery option in mine water [5].

As has been said before, in this research just the equilibrium of the TBP/IP6 in nitric acid with n-dodecane is going to be study.

2. Results and discussions

2.1 Propose system

The purpose of this work is to study an LLE-IE system to establish a new PUREX variant. Variants refer to change in some of the original conditions which in this case is adding a new molecule to the system. Original PUREX consist in TBP with HNO3 in a hydrocarbon. The proposed system consists in TBP with IP6 in solution con dodecane (**Figure 1**).

The IP6 presents 6 phosphates, it is water soluble and lightly soluble in ethanol and has a boiling point of 150 °C. The respective constants for calculations have been obtained from the literature [6].

The full chemical reaction with the purpose LLE-IE system is as present in **Figure 2**. It can be observed the interaction between the characteristic's actinides of

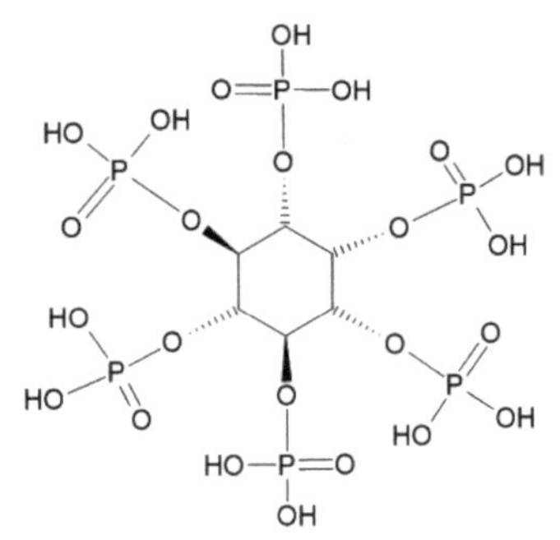

Inositol Polyphosphate
Molecular Formula: $C_6H_{18}O_{24}P_6$
Formula Weight: 660.035292
Composition: C(10.92%) H(2.75%) O(58.18%) P(28.16%)

Figure 1.
Inositol polyphosphate (IP6) molecule.

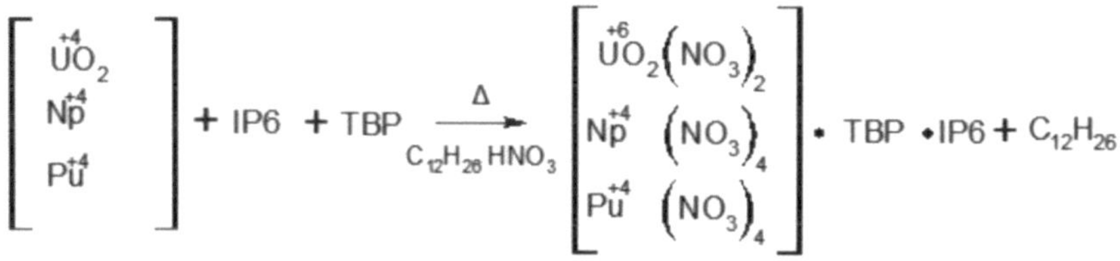

Figure 2.
Full extraction reaction presented for the purpose PUREX system.

a spent nuclear fuel and the TBP-IP6. In this reaction, the radiolitic effects are not considered.

2.2 Effects of water on the activity of TBP/IP6 in the H2O-dodecane system

The study system comprises 30% of TBP and 10% of IP6 (TBP/IP6) in solution with water and n-dodecane.

Considering that the distribution of water in the H2O–TBP/IP6–dodecane system be described using the Equation [7] (1)

$$x_1 = K_1\varphi_2 a_1 \exp\left(b_1\varphi_2^n\right) + K_2\varphi_3 a_1 \quad (1)$$

Where x_i, ai and φ_i are the molar fraction, activity, and volume fraction of the i component in solution respectively; indices 1, 2 and 3 refer to water, TBP/IP6 and dodecane respectively; in this work, by recommendation, we used n = 2.10; and the volume fractions of TBP/IP6 and dodecane were calculated neglecting water by:

$$\varphi_2 = \frac{x_2 V_2}{x_2 V_2 + x_3 V_3} \quad (2)$$

$$\varphi_3 = 1 - \varphi_2 \quad (3)$$

Table 1 present the value of x_1 calculated by Eq. (1). The following constants were used: $K1$ = 0.0795, $K2$ = 0.0029 and b_1 = 1.783 (used for dodecane too); V_2 = 273.8 cm^3 [1], v_3 = 228.6 cm3 [1], n = 2.10 and k_2 = 0.1.

a_1	x_1
1	0.044578274
0.9	0.040120447
0.8	0.035662619
0.7	0.031204792
0.6	0.026746965
0.5	0.022289137
0.4	0.01783131
0.3	0.013373482
0.2	0.008915655
0.1	0.004457827

Table 1.
Calculate mole fractions of water in TBP/IP6 solution with dodecane.

From Eq. (1) we can derive an equation for the molar coefficient of the activity of water.

$$f_1 = \frac{1}{\left[K_1\varphi_2 \exp\left(b_1\varphi_2^n\right) + K_2\varphi_3\right]} \tag{4}$$

$$lnf_1 = -\ln\left[K_1\varphi_2 \exp\left(b_1\varphi_2^n\right) + K_2\varphi_3\right] \tag{5}$$

The result of the Eq. (4) is a molar coefficient of $a_w f_1$ = *22.432452* and lnf_1 = *3.11050866*. To derive an equation for the molal coefficient of the activity of TBP, we used the cross-equation.

$$\left[\frac{\partial lnf_1}{\partial lnf_2}\right]_{m1} = \left[\frac{\partial lnf_2}{\partial lnf_1}\right]_{m2} \tag{6}$$

Where the derivatives with respect to the molar concentration m_2 y m_1 were calculated for constant m_1 and m_2 respectively. Differentiating (5), we obtain

$$\left[\frac{\delta lnf_1}{\delta m_2}\right] m1 = -\left\{\frac{\left[K_1 \exp\left(b_1\varphi_2^n\right) + K_1\varphi_2 \exp\left(b_1\varphi_2^n\right) n b_1 \varphi_2^{n-1} - K_2\right]}{\left[K_1\varphi_2 \exp\left(b_1\varphi_2^n\right) + K_2\varphi_3\right]}\right\}\left[\frac{\delta\varphi_2}{\delta m_2}\right] \tag{7}$$

The value m_2 can be calculated from the mole fractions of TBP/IP6 and dodecane,

$$m_2 = \left(\frac{x_2}{x_3}\right)\left(\frac{10^3}{M_3}\right) = \left(\frac{x_{20}}{x_{30}}\right)\left(\frac{10^3}{M_3}\right) \tag{8}$$

Where x_{20} and x_{30} are the mole fraction of TBP/IP6 and diluent in anhydrous solution; x_{30} = $1\text{-}x_{20}$; and M3 is the molecular mass of the solvent (170.33 g/mol). Then from (2), we obtain

$$\varphi_2 = \frac{V_2}{\left[\frac{V_2 + V_3}{\frac{x_{20}}{x_{30}}}\right]} = \frac{m_2 V_2}{\left[m_2 V_2 + \frac{V_3 x 10^3}{M_3}\right]} \tag{9}$$

From (9) we determinate the derivative $\delta\varphi 2/\ \delta m2$ for (7),

$$\frac{\partial\varphi_2}{\delta m_2} = \frac{V_2 V_3 x \frac{10^3}{M_3}}{\left[m_2 V_2 + \frac{V_3 x 10^3}{M_3}\right]^2} \tag{10}$$

Now, substituting the Eq. (10) in (7),

$$\frac{\partial\varphi_2}{\delta m_2} = \left\{\frac{V_2 V_3 x \frac{10^3}{M_3}}{\left[m_2 V_2 + \frac{V_3 x 10^3}{M_3}\right]^2}\right\} x \left\{\frac{K_1 \exp\left(b_1\varphi_2^n\right) + K_1\varphi_2 \exp\left(b_1\varphi_2^n\right) x\ n b_1 \varphi_2^{n-1} - K_2}{K_1\varphi_2 \exp\left(b_1\varphi_2^n\right) + K_2\varphi_3}\right\} \tag{11}$$

The right side of the Eq. (11) does not contain any value dependent on m_1. Then, integrating the Eq. (6), we obtain

$$lnf_2 = f_{20} - m_1 \left\{ \frac{V_2 V_3 x \frac{10^3}{M_3}}{\left[m_2 V_2 + \frac{V_3 x 10^3}{M_3} \right]^2} \right\} \left\{ \frac{K_1 \exp\left(b_1 \varphi_2^n\right) + K_1 \varphi_2 \exp\left(b_1 \varphi_2^n\right) x\, n b_1 \varphi_2^{n-1} - K_2}{K_1 \varphi_2 \exp\left(b_1 \varphi_2^n\right) + K_2 \varphi_3} \right\} \quad (12)$$

Where f_{20} is the TBP-IP6 activity coefficient in a binary (considering tri-n-butylphosphate and inositol hexaphosphate as one) anhydrous solution, which can be set at 1 in the first approximation. **Table 2** presents the results of the calculation by (12).

The deviations from the ideal values are moderate and increase with the activity of water and TBP/IP6 concentration.

2.3 Dissociation of nitric acid

Nitric acid is integral to the reprocessing of irradiated fuel and other LLE, the understandings its behavior is important. Nitric acid undergoes thermal and radiolytic degradation, the products of which include nitrous acid (HNO2) and nitrogen oxide species (NOX).

Eq. 13 shows the generic dissociation reaction of nitric acid.

$$HNO_3 \leftrightarrow H^+ + NO_3 \quad (13)$$

The equation for calculating the degree of dissociation is as follows:

$$K = \frac{[A^+][B^-]}{[AB]} = \frac{[C\alpha][C\alpha]}{C(1-\alpha)} \quad (14)$$

Where K is the equilibrium constant, AB is the reagent, A+ and B- ions (cation and anion respectively), C acid concentration and α dissociation degree. For alpha calculation purposes, we have an equilibrium constant of K = 2.598.

We will consider the dissociation of nitric acid using the polynomial Eq. (15), which has been adjusted from the data reported by [8]. In Eq. 15, the concentration of nitric acid [C] is in mol/dm3 and α the dissociation degree where α = 1 shows a complete dissociated acid and α = 1 a completely associated acid

a_1	m_1	Lnf_2	f_2
1	0.2438	0.05711146	0.94448879
0.9	0.2190	0.05115063	0.95013554
0.8	0.1945	0.04529659	0.95571399
0.7	0.1697	0.03940556	0.96136074
0.6	0.1449	0.03354904	0.96700749
0.5	0.1201	0.02772661	0.97265424
0.4	0.0963	0.02217066	0.9780733
0.3	0.0721	0.01655281	0.98358344
0.2	0.0474	0.01085124	0.98920742
0.1	0.0227	0.005182	0.9948314

Table 2.
Molalities of water m1 and TBP/IP6 activity coefficient f2 for a solution in n-dodecane.

$$\alpha = -2.64x10^{-6}[C]^4 + 2.6331x10^{-4}[C]^3 - 5.8558x10^{-3}[C]^2 - 1.54199x10^{-2}[C] + 1 \quad (15)$$

The following calculation describes the concentration of associated and dissociated nitric acid.

$$\left[NO_3^-\right] = \alpha \cdot [HNO_{3\ total}] \quad (16)$$

$$[HNO_3] = [HNO_{3\ total}] - \left[NO_3^-\right] \quad (17)$$

Where $[HNO_{3total}]$ is the sum of dissociated and associated nitric acid, $[NO_3^-]$ and $[HNO_3]$ are respectively the associated and dissociated acid concentration.

$$4HNO_3 \leftrightarrow 4NO_2^{\bullet} + 2H_2O + O_2 \quad (18)$$

It can be observed that after the 23 M the value increases again: due to the point of saturation of nitric acid and coexistence with non-associated species.

In nitric acid solutions, nitrogen oxide species, including HNO_2, NO_2 and NO, have been observed. The presence of these species in the absence of other reactants or radiation is attributed to the thermal decomposition of nitric acid. Non-dissociated nitric acid is thermally decomposed to produce $NO_2\bullet$ as shown in Eq. 18; notice that this reaction is non-elementary. This thermal decomposition of nitric acid in aqueous solution has been widely reported in the literature for different concentrations, high acidity and at high temperatures (**Table 3**).

2.4 Calculations of the equilibrium

The calculation method used in this research work is as follow:

1. The nitric acid and water activities are calculated from the data of [8].

2. The calculation of equilibrium implies the formation of the non-hydrated $HNO_3{\cdot}$TBP/IP6 monosolvate and the hydrated $HNO_3{\cdot}$2TBP/IP6 disolvate and $2HNO_3{\cdot}$TBP/IP6 semisolvate of nitric acid, and the equilibrium between them obeys the mass action law.

$$x_{ij} = \frac{K_{ij}a_a^i a_2^j}{f_{ij}} \quad (19)$$

where aa and a_2 are the nitric acid and TBP/IP6 activities, x_{ij} and f_{ij} are the molar fraction and rational activity coefficient of a solvate consisting of i acid molecules and j complex molecules (TBP/IP6). The parameter f_{ij} is calculated within the nonstoichiometric hydration concept by the equation

$$f_{ij} = \exp\left[h_{ij}(1 - a_1)\right] \quad (20)$$

where h_{ij} is the hydrate number of a solvate, and a_1 is the water activity.

3. The molar fraction of free water (nonbonded with solvates) is calculated by the equation

$$x_1 = K_1\varphi_1 a_1 \exp\left(b_1\varphi_1^n\right) + k_2\left[K_1\varphi_2 a_1 \exp\left(b_1\varphi_2^n\right)\right]^2 + K_2\varphi_3 a_1 \quad (21)$$

Eq. (21) is very similar to Eq. (1). As in (1), x_i, a_i, and ϕ_i are the molar fraction, activity, and volumetric fraction of the ith component in a solution. The volumetric TBP/IP6 and n-dodecane fractions are calculated without allowance for water as (2) and (3).

4. Organic phase nonideality is considered using the activity solvate coefficients calculated as

$$f_s = \exp\left[-b_2(1-\varphi_2)^{2.1}\right] \tag{22}$$

Molarity	α	NO_3^-	HNO_3
1	0.97898485	0.97898485	0.02101515
2	0.94780094	1.89560188	0.10439812
3	0.90793309	2.72379927	0.27620073
4	0.86080276	3.44321104	0.55678896
5	0.80776805	4.03884025	0.96115975
5	0.80776805	4.03884025	0.96115975
6	0.7501237	4.5007422	1.4992578
7	0.68910109	4.82370763	2.17629237
8	0.62586824	5.00694592	2.99305408
9	0.56152981	5.05376829	3.94623171
10	0.4971271	4.971271	5.028729
11	0.43363805	4.77001855	6.22998145
12	0.37197724	4.46372688	7.53627312
13	0.31299589	4.06894657	8.93105343
14	0.25748186	3.60474604	10.39525396
15	0.20615965	3.09239475	11.90760525
16	0.1596904	2.5550464	13.4449536
17	0.11867189	2.01742213	14.98257787
18	0.08363854	1.50549372	16.49450628
19	0.05506141	1.04616679	17.95383321
20	0.0333482	0.666964	19.333036
21	0.01884325	0.39570825	20.60429175
22	0.01182754	0.26020588	21.73979412
23	0.01251869	0.28792987	22.71207013
24	0.02107096	0.50570304	23.49429696
25	0.03757525	0.93938125	24.06061875
26	0.0620591	1.6135366	24.3864634
27	0.09448669	2.55114063	24.44885937
28	0.13475884	3.77324752	24.22675248
29	0.18271301	5.29867729	23.70132271
30	0.2381233	7.143699	22.856301

Table 3.
Calculation of values for the dissociation degree of nitric acid with to molarity in the solution.

5. The molar fraction x_i is determined as

$$x_i = \frac{c_i(1 - x_1)}{\sum c_j} \tag{23}$$

where the sum Σc_j is calculated for the first time as

$$\sum c_j = c_a + c_2 + c_3 \tag{24}$$

c_a, c_2, and c_d are the molar acid, TBP/IP6, and dodecane concentrations, respectively, and

$$c_2 = c_T – c_a \tag{25}$$

where c_T is the total complex (TBP/IP6) concentration in a solution, i.e., the formation of the monosolvate alone was initially assumed.

6. To calculate the molar fraction of free complex x_{2f}, we write the equation

$$\begin{aligned} & x_1 + x_{2f} + x_3 \\ & + \left\{K_{11}a_a x_{2f} f_2 + K_{21}a_a^2 x_{2f} f_2 \exp\left[h_{21}(a_1 - 1)\right] + K_{12}a_a x_{2f}^2 f_2^2 exp\left[h_{12}(a_1 - 1)\right]\right\} \\ & * exp\left[-b_2(1 - \varphi_2)^{2.1}\right] = 1 \end{aligned} \tag{26}$$

7. The value of x_{2f} calculated by the Eq. (26) is used to determine the molar fractions x_{ij}. The molar concentrations c_{ij} are then estimated by the equations

$$c_j = \frac{x_j d * 1000}{\Sigma x_i M_i} \tag{27}$$

Parameter	Value	Units
% TBP	30.00%	%
% Dodecane	60.00%	%
% IP6	10.00%	%
Molarity HNO_3 [M]	9	mol/L
Water activity [a_w]	0.6	
Molecular weight HNO_3	63.01	g/mol
Molecular weight Dodecane	170.34	g/mol
Molecular weight TBP	266.29	g/mol
Molecular weight IP6	660.04	g/mol
ρ HNO_3	1.5129	g/cm^3
ρ Dodecane [d_0]	0.73526	g/cm^3
ρ TBP	0.973	g/cm^3
ρ IP6	1.3	g/cm^3
Acid concentration [ca]	9	mol/dm^3

Table 4.
Principal input parameters and its values.

Parameter	Value
d complex TBP/IP6	1.06984
HNO_3 Dissociation degree [α]	0.5615298
Volumetric fraction of complex [ϕ2]	0.4
Volumetric fraction of dodecane [ϕ3]	0.6
Molar fraction of water [x_1]	0.0257969
Molar activity coefficient water [f_1]	0.9583666
Solvate molar activity coefficient [f_s] Organic phase	0.4245719
Complex molar activity coefficient [f_2]	0.96700749

Table 5.
Principal results for the equilibrium calculation with 30% TBP/10% IP6 in.

where d is the density of a solution, and x_i and M_i are the molar fraction and mass of the ith component.

The values of c_j are used to correct the molar fractions in compliance with Eqs. (23) and (24).

The calculated acid molar concentration c_{ac} is further found as (28) and the calculated complex molar concentration c_{tc} is estimated as (29)

$$c_{ac} = c_{11} + c_{12} + 2c_{21} \tag{28}$$

$$c_{tc} = c_{2f} + c_{11} + 2c_{12} + c_{21} \tag{29}$$

Table 4 presents all the principal input parameters. The values presented in the table are the one who has been used to solve the equilibrium equations.

The concentration of the acid allowed to know the activity of water in the system, which have a value of 0.6 which represents a large amount of water to form the aqueous phase, since a water activity value equal to 1 would represent that we have the total disposition of water to hydrate.

Table 5 presents the results of the calculation in the equilibrium.

3. Conclusions

As first step in the overall objective of the study of the equilibrium in the LLE-IE, the kinetic data and constants values has been investigated to produce an initial dynamic model of the interaction of the TBP/IP6 in aqueous conditions. The effects of water in the activity of the TBP/IP6 has been evaluated. As it can be seen, the deviations from the ideal values of the molar coefficient of the system TBP/IP6 $f2$ are moderate and increase with the activity of water and TBP/IP6 concentration. The density of the complex makes precipitation possible and enough availability of dissociated acid makes this complex suitable for redox reactions.

Author details

Munoz Ayala Israel[1*] and Vera Roberto Carlos[2]

1 National Polytechnic Institute, Mexico City, Mexico

2 Research Institute of the Faculty of Medicine: "Tomas Frías" Autonomous University, Potosí, Bolivia

*Address all correspondence to: israel.muoz07@gmail.com

References

[1] Davis, W., & Mrochek, J. (1966). Extraction Chemistry. Proc. of Intern. Conf (p. 152). Geteborg, Sweden: A. A. Pushkov.

[2] Ramanujam, A. (1998). An introduction to the Purex Process. IANCAS Bulletin, 14(2), 11–26.

[3] Berk, Z. (2013). Food Process Engineering and Technology (Second Edition). Advances in Molecular Toxicology.

[4] Xintao, W., Qui, L., Hongsen, Z., Jingyuan, L., Rongrong, C., Rumin, L., . . . Jun, W. (2017). Rapid and Efficient Uranium(VI) Capture by Phytic Acid/ Polyaniline/FeOOH Composites. Journal of Colloid and Interface Science, doi: http://dx.doi.org/10.1016/j.jcis.2017.09.054.

[5] Paterson-Beedle, M., Readman, J., Hriljac, J., & Macaskie, L. (2017). Biorecovery of uranium from aqueous solutions at the expense of phytic acid. Hydrometallurgy, 524–528. doi:104. 524–528. 10.1016/j.hydromet.2010.01.019.

[6] Muñoz Ayala, I. (2021). Captura selectiva de actínidos en ensambles de combustible nuclear gastados de un BWR. Ph.D. Thesis. Instituto Politécnico Nacional, Mexico City.

[7] Gladilov, D. Y., Nekhaevskii, S. Y., & Ochkin, A. V. (2006). A thermodynamic description of the distribution of water in H2O-tributyl phosphate and H2O-tributyl phosphate-solvent systems. Russian Journal of Physics and Chemistry, 80(12), 1934–1939. doi:10.1134/S0036024406120120

[8] Davis, W. J., & Bruin, J. (1964). New activity coefficients of 0-100 per cent aqueous nitric acid. J. Inorg. Nucl. Chem., 26, 1069–1083.